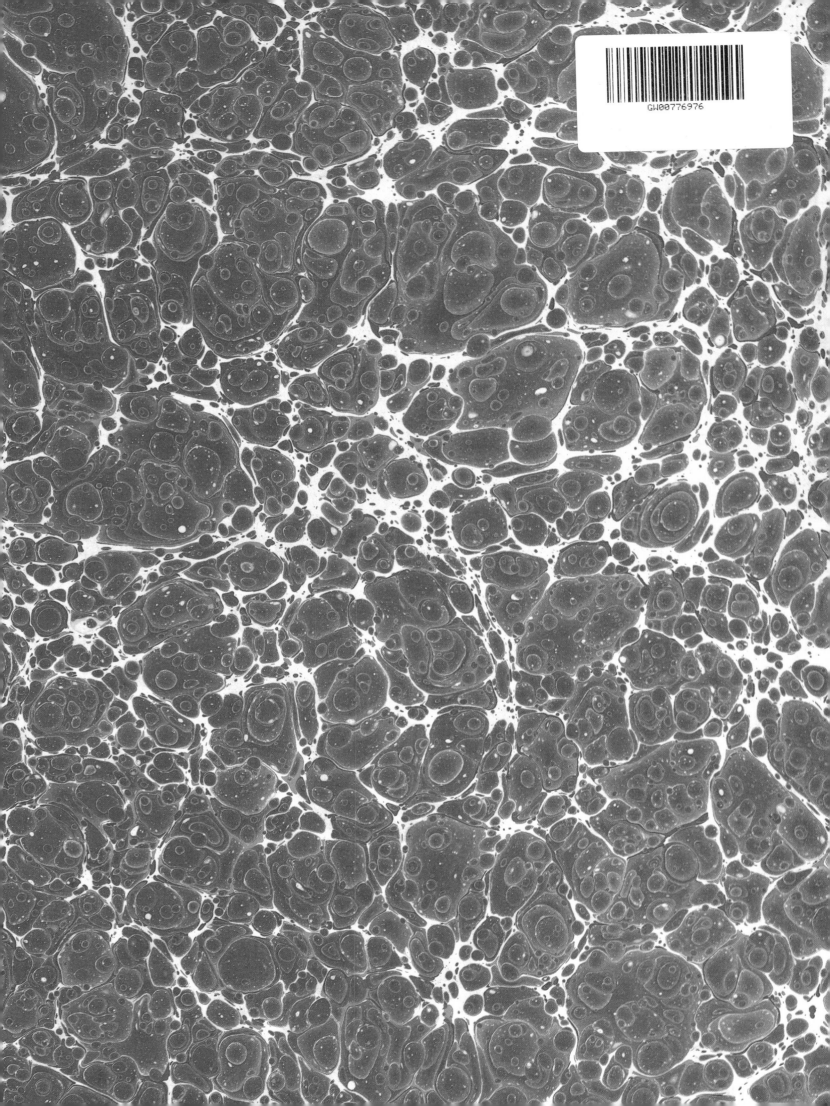

Black Forest Clocks

Black Forest Clocks

Rick Ortenburger

1469 Morstein Road, West Chester, Pennsylvania 19380

This book is dedicated
to my mother Marie Anna Ortenburger,
my grandmother Emily Hackstock,
and my wife Elfi Ortenburger.

Copyright © 1991 by Rick Ortenburger.
Library of Congress Catalog Number: 90-63797.

All rights reserved. No part of this work may be reproduced or used in any forms or by any means—graphic, electronic or mechanical, including photocopying or information storage and retrieval systems—without written permission from the copyright holder.

Printed in the United States of America.
ISBN: 0-88740-300-X

We are interested in hearing from authors with book ideas on related topics.

Published by Schiffer Publishing, Ltd.
1469 Morstein Road
West Chester, Pennsylvania 19380
Please write for a free catalog.
This book may be purchased from the publisher.
Please include $2.00 postage.
Try your bookstore first.

Front cover photo:
Trumpeter von Säckingen by Emilian Wehrle, Furtwangen. (Photo by Steve Ridnor, Photographic Images, Agoura, CA.

Back cover photos:
Picture frame cuckoo, carved cuckoo and early wood wheel clock. (Deutsches Uhrenmuseum, Furtwangen).

Contents

	Foreword	7
	Acknowledgements & Thanks	9
Chapter 1	**The History of Black Forest Clocks: From Cottage to Factory**	11
	The Beginnings of Specialized Labor	15
	Changing Technologies	19
	The Clock Dealer and Travelers	22
	From Small Clockmakers to Factory: American System	30
	The Woodcarving School	34
Chapter 2	**The Cuckoo Clock**	53
	Cuckoo and Quail Birds	87
	Hands	91
	Weights and Pendulums	91
	Wood and Brass Plate Cuckoo Movements	95
Chapter 3	**Johann Baptist Beha of Eisenbach**	100
Chapter 4	**The Clockmakers of Furtwangen**	123
	A. Mayer, Uhrenfabrik Schönenbach bei Furtwangen	123
	Lorenz Furtwängler Söhne (LFS)	123
	Badische Uhrenfabrik	124
	Gordon Hettich Sohn Uhrenfabrik	127
	Union Clock Co.	129
	Lorenz Bob	130
Chapter 5	**The Clockmakers of Triberg and St. Georgen**	132
	Jahreuhren-Fabrik, GmbH, August Schatz & Sohne	132
	Phillip Haas and Söhne	132
	Matthias Bäuerle GmbH	132
	Tobias Baeuerle	135
	Andreas Maier	135
Chapter 6	**The Villingen Clockmakers**	137
	C. Werner, Uhrenfabrikation	137
	Gebrüder Wilde Uhrenfabrik	137
	August Noll	139
	Uhrenfabrik Villingen Aktiengesellschaft and Uhrenfabrik, Villingen J. Kaiser, GmbH	141

Chapter 7	**The Schwenningen and Schramberg Clockmakers**	143
	Friedrich Mauthe	143
	Jacob Kienzle	145
	Gebrüder Junghans	147
	Hamburg American Clock Company	157
Chapter 8	**The Clockmakers of Neustadt, Lenzkirch, and Schonwald**	160
	Uhrenfabrik Fürderer Jaegler & Cie and Uhrenfabrik Neustadt i. Schw. Aktiengesselschaft (Corp.)	160
	Winterhalder and Hofmeier	160
	Aktiengesellschaft Für Uhrenfabrikation	166
	Karl Joseph Dold Söhne Uhrenfabrik	175
	Wehrle Uhrenfabrik GmbH	175
Chapter 9	**Trumpeter Clocks**	198
	Jacob Bäuerle	198
	Emilian Wehrle, Furtwangen-Schönenbach	198
Chapter 10	**Small Clocks and Their Makers**	229
	Jakob Herbstreith	229
	Josef Sorg	231
Chapter 11	**Organ and Flute Clocks and Their Makers**	233
	Ignatz Bruder	233
	Ruth & Söhne	234
Chapter 12	**Animation Clocks**	236
Chapter 13	**Black Forest Regulators**	246
Appendix 1	**Reproduction of pages from the 1904 Twelfth Annual Catalog, St. Louis Clock and Silverware Company.**	251
Appendix 2	**Carved Clocks with French Style Movements**	256
Appendix 3	**The Black Forest Clock Industry 1875-1892: From A Horological Magazine's Viewpoint**	259
	Wanted by Author	263
	Bibliography	264
	Index	265

Foreword

My interest in old clocks began in 1972 when my wife's aunt gave me an old wall clock that she had bought at the time of her wedding in 1928. My interest in clocks grew and I began to appreciate the beauty of the so-called Vienna Regulator. Through an ad in a newspaper, my father found a man who had collected over 2000 clocks and watches. He suggested that I give this man a call. From there I joined the National Association of Watch and Clock Collectors and, among other things, was introduced to a wealth of information about clocks, encountered a variety of clocks, and met other collectors by attending their conventions.

I had always had an interest in Black Forest clocks. I had been in the Schwarzwald while I was stationed in Germany. Being full of rich farmland and forest, it was a beautiful area of the world. Why did this area and its clocks intrigue me so much? I cannot explain it, but perhaps it was because my grandfather, William Ortenburger, was a wood turner at a mill near Oberschlesien prior to his coming to America in 1907. And maybe it was because many Black Forest clocks do much more than tell time; some have the cuckoo, the quail, music, trumpeter movements, or animation, etc. There is something wonderful about the beauty of a Lenzkirch factory clock, with its brass decorated case and highly accurate, reliable movement.

Around 1981 I had the idea to write a book on Black Forest clocks and started to photograph all the clocks I could find that related to this subject. The book took me to New York, New Jersey, Massachusetts, Texas, Arkansas, Oregon, Ohio, California, and Germany. I met many collectors of German clocks and many helped with this book (see the Acknowledgements and Thanks). The book finally seemed a reality when I met my publisher, Peter Schiffer. He expressed an interest in the book and I thank him for listening and publishing this book.

The most well-known clock factories in the Black Forest are included in this book. Considering the number of small clockmaking enterprises at the time, it is difficult to place them in categories. And because of the limited size of this book, it is impossible to include all of them. The factories that shipped clocks to the U.S.A. were especially considered for inclusion to this book. Because the clocks were being shipped in large containers across the U.S. by antique importers, clock collectors in the U.S.A. could find them in droves in the 1950s to the mid-1970s.

Only what is known from previous writings about the factories along with this writer's opinion is included. The order in which the factories or clockmaker is listed does not represent a ranking order. The history of these clockmakers is intended to give the reader a general insight into clockmaking in the Black Forest, which still continues today though in a limited way and to a much smaller degree.

I wish to thank the previous writers on Black Forest clocks. Without their work, this book would have been incomplete. They are E. John Tyler (England), Gerd Bender, Berthold Schaaf, Wilhelm Schneider, R. Mühe, H. Kahlert, H. Jüttermann, (Germany) Karl Kochmann, and Dana Blackwell (U.S.A.).

I would also thank Elfi Ortenburger for her endless hours of translation of German Black Forest history into English which helped to tell this story in a more complete manner. The people who added so much to the creation of this book are all listed in the Acknowledgements and Thanks. No matter what these people contributed, whether small or large portions, they are all to be commended for coming forth and offering what they had, which in the end added something to the history of the Black Forest clockmakers by making this book possible. Thank you all!

All clocks pictured in this book are in private collections and are photographed by the author, unless otherwise noted.

I sincerely hope that you will enjoy this book. Your comments and correspondence are welcome.

Rick Ortenburger

The Trumpeter's Song
by Viktor von Scheffel

Life is arranged unpleasantly, (literally: in an ugly way)
That thorns are next to roses,
And whatever the poor heart longs for and puts into rhyme,
At the end is parting:
In your eyes I once have read,
In them a light was sparkling with love and happiness:
The Lord be with you! It would have been too beautiful,
The Lord be with you, it was not meant to be!-

Sorrow, envy, and hate, I, too, have felt,
A storm-proven, wearysome wanderer.
I dreamed of peace, then, and of quiet hours,
When my path led up to you.
In your arms I wanted to recover,
Out of gratitude devote my young life to you.
The Lord be with you! It would have been too beautiful,
The Lord be with you, it was not meant to be!

The clouds are moving across the sky,
A rain shower passes over pastures and fields,
For parting, just the right weather,
Grey like the sky, the world is standing before me.
Yet, may there come good or bad,
You slender maid, in faith, I will remember you!
The Lord be with you! It would have been too beautiful,
The Lord be with you, it was not meant to be!-

Acknowledgments and Thanks

The following people helped with this book in one way or another and I thank them at this time. Many of them opened their collections for photography for inclusion in this book. Many shared their knowledge. Without them, this book would not be in print today.

Thanks to: Dennis Holt, Oregon; Bud Saiben, California; La Mar Rozsa, California; George Edwards, Pennsylvania; Amy Smith, Pennsylvania; Joseph Schumacher, Delaware; Barry Malkemes, Pennsylvania; Bill Heron, Pennsylvania; Junie Fleming, Jr., North Carolina; Richard Centofanti, California; Royal Color, Inc., Ohio; Clocks of Distinction, George Hamburger, Hunters Hill, Australia; H. Stanley Bourne, Virginia; Beatrice Techen and the Deutches Uhrenmuseum, Furtwangen-Schwarzwald, Germany; Jim Carlisle, Arkansas; Ilse Kaiser and Verlag Müller, Villingen, Schwarzwald, Germany; Gerd Bender, Germany; Karl Kochmann, Antique Clocks Publishing, California; Berthold Schaaf, Germany; Dr. Wilhelm Schneider, Germany; C. Pfeiffer-Belli and Verlag Georg D.W. Callwey, München, Germany; Robert Spence, Washington; Horological Data Bank, NAWCC Museum, Pennsylvania.

Also thanks to: Dennis Ortenburger, California (for encouraging me to write about clocks in books and articles); Douglas Barr, Ohio; Frank Snyder, New Jersey; Eric Fuchslocher, California; Adolf Haas, New York; Fred Bausch, California; Bob MacIver, California; Robert Zimmerman, California; Eugene Kramer, Texas; Ron Holman, California; Phil Rasch, Texas; Euro Am Clocks, California; Antique Clock Gallery, California; Harry Larson, California; Joe Chesatis, New Jersey; Rudi Kemper, Ohio; Danny Buchanan/Judy Schriver, Pennsylvania; Pete and Barbara Eggimann, Massachusetts; Peter Mikkelsen, California; John Roach, California; Royce Hulsey, California; Art Bjornestad, California; Joe Lyons, Lyons Antique Clocks, California; Corliss Lee, California; Roy Irick, California; Lena Kelner, California; Andy Copeland, Texas; Ivan Godwan, California; Fast Foto/Avnon Film Lab, California; Patti & Joel Cohen, California; Steve Kukich, California; Sundial Farm, Larry & Maria Thompson, New York; NAWCC 1980 San Jose Regional Exhibit personnel, California; NAWCC 1986 Cleveland National Exhibit personnel, Ohio; Elfi Ortenburger, California; Peter Schiffer, Pennsylvania; Frank Friedrich, Maryland.

If anyone was inadvertently forgotten, I would like to take the opportunity to thank you now.

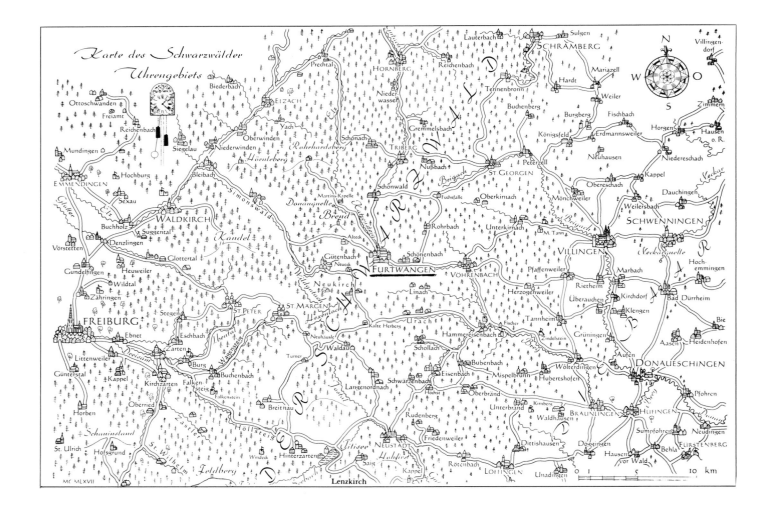

Historical map of the Black Forest Clockmaking region. (Permission to reprint this map was granted by Verlag Georg D.W. Callwey, München, Germany.

CHAPTER 1

The History of Black Forest Clocks: From Cottage to Factory

The Schwarzwald or "Black Forest" is located in the southwest of Germany. It is about 90 miles long by 30 miles wide and runs almost parallel with the Rhine River. It is very mountainous and rich with timber. Though there is much concern now about the effect of acid rain on this extremely beautiful area of the world, at one time the trees in this area were extremely close together and very thick. When one walked through the forest it was very dark, due to the denseness of the trees. Thus the name "Black Forest."

While principally an agricultural area, an alternative to farming was born in the mid-seventeenth century. This occupation was clockmaking. It was first performed on a part-time basis, usually during the winter months. Summers were very short and farming used most of the inhabitants' time and energy during this season. In the harsh winters, however, the families had to remain indoors. There was a desire to fill idle time during these winter months with something useful and perhaps profitable.

Clockmaking fit the need and soon became a necessity for economic survival in this isolated area. In part this was due to a tradition that the family farm was always willed to the oldest son. Other members of the family were dependent on the older brother or were forced to find work elsewhere. They turned to other forms of employment, including woodcarving and the glassware industry. Woodcarving was used extensively in later years, as this book will show.

Legend has it that the Black Forest clock was introduced to the area in 1640, when a travelling glassware salesman brought a wood wheel clock back from Bohemia. The Black Forest farmers copied the clock for decoration in their houses.

More scholarly sources on Black Forest clocks suggest other dates around 1667 or 1685, and Lorenz Frey from St. Märgen is credited with making a simple clock around 1690. In the town of Waldau, there is a tablet situated on a building known as Glashof. It states that the brothers Kreutye made the first Black Forest clock there in 1640. It should be stated that a 200 year anniversary festival was to have been held in August, 1870 in Waldau, which was to celebrate the beginning of the Black Forest Clock industry. It was cancelled due to the outbreak of the Franco-Prussian War in July, 1870. This would indicate that the industry started around 1670.

Some of the early clockmakers were Simon Dilger of Schottenbach (born 1672), who became one of the first independent masters, Franz Ketterer of Schönwald (born 1673), Johann Dufner of Schönwald (born 1673), and Mathias Löffler of Gutenbach (born 1680). The clocks were made with wooden frames and gears. They were primitive and driven by weights made of stones usually taken from a river because they had a polished look and seemed suitable for the clock. As the industry grew in the Furtwangen and Neustadt area and surrounding villages, so did the number of clock masters. They totalled thirty-one in 1740.

The tools used to make these early clocks were simple: a pair of compasses, drills, a saw, and a knife. A wheel cutting machine was invented by Mathias Löffler around 1725, though a similar machine may have been also introduced by Friedrich Dilger, son of Simon Dilger. The machine for making pinions was made by Georg Wilmann of Neustadt around 1740. The inventions of these machines cut the time of making clocks from six days to one and led to the addition of striking trains or gears in the movements.[1]

Clocks with moving figures could be found in the early years of German clockmaking. Some of the early wood wheel clocks had carved wood

Above: The Black Forest in winter, Feldberg. The highest peak is 1493 meters, or 4900 feet, above sea level. (Antique Clocks Publishing Archive.)

Below: Another view of the Black Forest, the city of Schramberg, 800 meters above sea level from Hardt. (Antique Clocks Publishing Archive.)

Black Forest flute clock, made about 1780, with 32 pipes made from wood, and 7 songs. Heavily carved baroque shield painted gold and carved by Matthias Faller of Neukirch. (Gerd Bender, *Die Uhrmacher des hohen Schwarzwaldes und ihre Werke, Bank I*, Verlag Müller, Deutsches Uhrenmuseum, Furtwangen/Schwarzwald, Germany.

figures on the top of them that moved when the clock struck its metal or glass bell. Most of these clocks found today are in museums or private collections in Europe. This style of clock developed the talent of the clockmaker and many let their imaginations take over in the making of these clocks.

As early as the 14th century, "art clocks" with automated figures were made in Europe. They were considered technical masterpieces and gained world renown. They were referred to as "figure clocks," not automated clocks. Johann Friedrich Dilger from Urach was widely known for his creation of figure clocks around 1730. Matthäus Hummel from Jägersteig, near Waldau, was also well known for his technological abilities in the making of figure clocks. It appears that the Black Forest clockmakers were also inspired by the tower clocks that they saw during their travels. Many of the tower clocks in Europe have moving figures that appear when the hour is struck.

In the 19th century, Michael Dorer (1811-1853) was a leader in making automated clocks near Furtwangen. He learned the clock trade from his father, Michael Dorer (1773-1839), who was also a maker of this type of clock.[2]

The Kuhschwanz, or "cow tail," pendulum was attributed to Christian Wehrle of Simonswald around 1740. The short pendulum swayed in front of the dial at a very fast pace. This appropriate nickname probably reflects the farming occupations of the people who invented the pendulum. Shortly afterward, the long pendulum and anchor escapement became part of the clock.[3]

The invention of the cuckoo clock has been attributed to Franz Anton Ketterer of Schönwald around 1730. However, it is now known that an earlier iron wheel cuckoo with automated figures existed prior to this date. Also credited with the development of the cuckoo were Josef Kammerer of Furtwangen, Josef Ganter of Neukirch, and Michael Dilger.[4]

14 Black Forest Clocks

The dial of a flute hallclock made by Ignaz Bruder in Simonswald. (For complete clock see page 235) (Gerd Bender, *Die Uhrmacher des hohen Schwarzwaldes und ihre Werke, Band I,* Verlag Müller, private collection, E.I. Amrein, Basel/Schweiz).

Black Forest shield animation clock with four moving figures that depict the beheading of John the Baptist. The clothes of the figures are from the period of about 1800, when the clock was made. (Gerd Bender, *Die Uhrmacher des hohen Schwarzwaldes und ihre Werke, Band I,* Verlag Müller, Badisches Landesmuseum, Karlsruhe, Germany.)

Whatever its origins, the cuckoo clock was the most famous of the Black Forest clocks and today it continues to be a very popular export. So widespread is its fame that the cuckoo has become synonymous with the Black Forest clock, or "Schwarzwalduhr." However, if one visits the Historical Clock Museum in Furtwangen, Black Forest, it becomes clear that the cuckoo clock was only a small part of the Black Forest clock industry.

Every writing on the cuckoo clock up to the mid-1980s has given credit to the Black Forest clockmaker as its originator. Bender reports that a man named Streyrer, who was a chronological event expert, asserted that the cuckoo clock did not originate in the Black Forest. No one really wished to believe this, even though most agreed with him that the wooden wheel clock also did not come from the Black Forest. His belief about the cuckoo clock was not totally clear, however, so no one believed his version.

A Father Jäck from the town of Gütenbach had less professional knowledge than Streyrer and his version about the cuckoo clock left no doubt that the cuckoo clock did originate from the Black Forest. Father Jäck was not as knowledgeable as Streyrer, but since he was a native of the area people tended to believe his version about Anton Ketterer of Schönwald being the cuckoo clock inventor.

The earliest known wooden wheel clock using a calling bird was possibly copied from an iron wheel automation clock. Wood was used, probably because it was more readily available and less expensive than iron. This particular clock has a rooster figure and the call of the rooster. One of these clocks is in Castle Rheinsberg. While there are many opinions about the relation of this rooster clock to the cuckoo, and no solid proof that the cuckoo was patterned upon it, it is clear that the rooster call clock was made first.

There is an iron wheel cuckoo clock (which the author has seen) in a private collection with an original wood carved cuckoo bird with a moving beak and two restored bellows. A painted wood figure above that moves one hand and the pendulum sways in front of the dial. The figure's right arm is made from iron and hits a bronze bell on top. The left hand looks like it had something in it, but it is broken. There was also a cuckoo door in front, but it is missing. There is a lot of similarity between the ways the early wood wheel cuckoo clocks and this iron wheel cuckoo were made. This iron wheel cuckoo clock is also similar to the iron wheel clocks that were made in the Nurnberg/Augsburg region. Alternatively, it could also be either Tyrolian or Italian as it is similar to the automated clocks that came from Venice, Italy.

It appears that it was made at the beginning of the 17th century, but certain parts and the pendulum indicate it was probably made between 1657 and 1676.

The conclusion of Dr. Wilhelm Schneider of Regendorf, Germany, who has done extensive research on this iron wheel cuckoo, is that the Black Forest wooden wheel cuckoo clock was not the first clock to have the cuckoo mechanism.[5] There seems to be a connection between cuckoo mechanisms and iron wheel clocks made in the 17th century in the southern German regions. These clocks from Augsburg and Nurnburg were influenced in turn by the Arabic and Greek water clocks. They had used animals, trumpeters, and whistling and singing birds, though not the cuckoo. You can form your own opinion on this version, but the fact is this iron wheel cuckoo from around 1660 does exist!

THE BEGINNINGS OF SPECIALIZED LABOR

There were many workers involved in the making of a clock. Some made the wooden frames, finished the dials, and worked with the tools that were a necessity to clockmaking. Others made the hands, painted the dials, or turned or carved wood. The clockmaker cut teeth in the wheels and assembled the movement.

In the early years of Black Forest clockmaking, each clockmaster usually had one journeyman and one apprentice. The apprentice usually worked for two to four years, and the journeyman was adding to his skill and understanding in the hopes of eventually having his own clock business. Work days were very long. A small shop would produce perhaps four clocks per week. At the end of the 18th century, there were about 1,000 clockmakers supported by about 300 workers and they produced about 200,000 clocks per year.

Competition spread as more journeymen became clockmakers and opened their own shops. The clockmakers produced larger quantities of clocks, but often lost quality in the process. The clocks they made were not very reliable and inferior to the earlier ones. By 1840 more clock shops were established, usually specializing in a certain type or model of clock. A clock master, his journeyman, and apprentice could produce 18 twelve-hour clocks, 14 twenty-four-hour clocks, 12 Schottenuhren, or 7 eight-day clocks in about a week. Shortly after this time, the American clock industry began to have an impact on what the Black Forest clockmaker sold to England. The entry of the Americans into the industry definitely slowed down the production of clocks in Black Forest, and many clockmakers left the region for the U.S.A.

16 Black Forest Clocks

Black Forest shield animation clock with the figure of a watchman or soldier in a barracks with two towers. In between those two towers the soldier moves back and forth. Made by Michael Dorer in Furtwangen about 1840. (Gerd Bender, *Die Uhrmacher des hohen Schwarzwaldes und ihre Werke, Band I*, Verlag Müller, Deutsches Uhrenmuseum, Furtwangen/Schwarzwald, Germany.)

Black Forest flute clock with 3 moving figures by Jakob Streifer in Furtwangen, made about 1830. (B. Schaaf, *Schwarzwalduhren*.)

Wooden Wheel Cuckoo clock made around 1760. The face is paper and painted with water colors, with bell strike. (Deutsches Uhrenmuseum, Furtwangen, Germany)

Wood wheel clock with animated figure and strike on a glass bell, made around 1700 (Historical Clock Museum, Furtwangen, Black Forest)

Black Forest astronomical clock. (Deutsches Uhrenmuseum, Furtwangen, German)

18 Black Forest Clocks

Flute clock with carved shield and figures that turn back and forth and also in a circle. The black bird also moves back and forth. Six songs. Made around the end of the 18th century. (B. Schaaf, Schwarzwalduhren.)

Bahnhäusle, or fretwork case, Black Forest clock. The painting shows a group of workers toasting a count. The toast is written under the painting: "To the count and his house." (Gerd Bender, *Die Uhrmacher, des hohen Schwarzwaldes und ihre Werke, Band I*, Verlag Müller, Deutsches Uhrenmuseum, Furtwangen/Schwarzwald, Germany.)

CHANGING TECHNOLOGIES

The first Black Forest clock movements contained wood gears and arbors. The arbors had wire inserted in their ends as pivots, which were inserted in holes in the wooden frame, or plates, of the movement. Wood arbors were still used around 1840, but their use decreased until they were discontinued totally, probably around 1870.

Originally, the gears were completely made of wood. Escapement gears were eventually made from brass and the castings were first imported from Solothurn and later Nurnberg. With the introduction of brass casting into the Black Forest clockmaking industry, other wheels were made from brass in addition to the escapement wheel.

Wood frame movements were still used in the late 1800s. The frame makers gathered large supplies of beech wood, some of it being imported from Freiburg and the Rhine Valley. It could be found drying in large stacks outside of the large farm houses or cottages. The frames were made by the clockmaker who worked on six or seven at a time, marking the holes for the gears, drilling them and inserting the brass bushing. The gears came unfinished from the foundry and were smoothed on a lathe by a worker or the clockmaker. The teeth for the gears were cut by the clockmaker on a cutting machine known as a Zahnstuhl.

After the teeth were cut, the gears were mounted on the arbors which already had the pinion inserted. When all gears were finished, they were put in the clock frame and tested for accuracy. The clock had one hand and the running duration of the movement was 12 to 15 hours. They were not very accurate, perhaps losing or gaining 20 minutes per day.

The clockmaker usually specialized in making one type of clock. He sometimes stamped the front of the clock frame with his initials, which was hidden by the dial. Others wrote their signature on the backboard of the movement, but these usually faded with time and are usually indecipherable today. For this reason, little is known about the clockmakers and it can be difficult to identify a clock's maker. The clock movement was protected from dust by side doors made of thin fir wood, but many of these are missing from the clocks found today.

The first Black Forest clock dials consisted simply of a chapter ring with decoration above and below it, and were made from a single piece of wood and crudely painted. Later, as the movement became larger and more elaborate, the dial covered the whole movement and sometimes was painted with water color. Occasionally they were printed on paper and applied to the wood and varnished

Wall hanging cuckoo clock made by Johann Beha in Eisenbach about 1860. The oil painting is on tin and depicts the birth of Christ. The wood plate, brass gear movement will run for one day. Enamel dial. The case is also inlaid. Brass weights and pendulum bob. (Photo provided by Dr. Wilhelm Schneider, Regendorf, Germany.)

over or covered by a sheet of glass. The idea of a paper dial is believed to have originated from Mathias Grieshaber of Gutenbach. Early dials sometimes had a smaller quarter hour dial below the main dial.

About 1780 the standard Black Forest style of dials (shield type) came about. They consisted of a square with a semi-circle above it. They were made from planks of wood, planed smooth, and sawed by a water-powered saw. Another circular layer of wood was glued to the base and the dial was then turned on a lathe to make the chapter ring convex. The extra piece of wood helped to stabilize the dial. The dial turner then passed the dial to the painter who chalked it. When dry, the surface was smoothed out with tripoli powder and pumice on a turntable.

20 Black Forest Clocks

Wood wheel clock with reverse glass painting, with strike on glass bells. Dated 1777. (Duetsches Uhrenmuseum, Furtwangen, Germany)

Wood wheel clock with two hour dials and a quarter dial. Dated 1772 (Deutsches Uhrenmuseum, Furtwangen, Germany)

Wood wheel musical clock with the strike on 10, glass bells. (Deutsches Uhrenmuseum, Furtwangen, Germany)

A wood wheel Cuckoo clock made about 1800 with Schottenuhr movement. (Deutsches Uhrenmuseum, Furtwangen, Germany).

Wood wheel clock with glass bell for strike, made in 1789 (Deutsches Uhrenmuseum, Furtwangen, Germany).

Wall hanging cuckoo clock with 30-hour wood plate, brass gear movement. Circa 1860. (Photograph from slide bought at Deutsches Uhrenmuseum, Furtwangen.)

Animation shield clock with the figure of a Turk with moving eyes and mouth. Half-hour strike by Bertold Schyle, Schonach about 1900. (Deutsches Uhrenmuseum, Furtwangen/Schwarzwald, photo Callwey Verlag, München, Germany.)

The dial was then painted with a white background and allowed to dry. Flowers or other decorations were then added. Finally, the dial was varnished. Oil paint for dials was believed to have been introduced in 1770 by Mathias Dilger and varnish by Georg Gfell (Urach) in 1775. Later, transfers were used on dials as painting them became too expensive. The shield dial was made well into the 19th century. The designs were varied; the French preferred flowers on the dial, Germans preferred small paintings in the corners, and the Baltic countries preferred the designs to have columns in them. The cuckoo clock also used this shield type dial.

The bells used for the strike on the clocks were originally made of glass. Shaped like wine glasses with ball feet, they were hung upside down in a wood frame above the movement. Sometimes multiple glass bells were incorporated to play music.

The metal bell came later and was used extensively on these clocks as time went on. Around 1760 Paul Kreuz of Waldau began casting the metal bells, which had earlier been bought from Solothurn or Nurnberg.

Wire gongs were also used with striking clocks and the first were made by Carl Dold in Furtwangen around 1830. Later they were also made in Schwenningen.

On the earliest clocks the weights were hung on cords with a counterweight on one end to keep the cord on the pulley. Later the cords were replaced with chains. The chainmaking machine is credited to Augustin Kienzler of Freiburg in 1837, and was foot operated. Christian Burger of Gütenbach made a better machine in 1858, but these machines were made late in relation to the traditional Black Forest movement.

The size of the traditional Black Forest movement was normally about 6¾ inches x 5½ inches x 4¾ inches. It should be mentioned that around 1760 the Schottenuhr movement was made by J. Dilger. The name came about because J. Dilger worked in the Schottenhof near Neustadt. These movements measured about 4 1/3 inches x 3¼ inches x 3¼ inches. The dials were also smaller, about 6 inches in length, and the pendulum length was about 11¾ inches. This movement was used by many of the factories in later years. A Phillip Haas factory page in this book shows that this firm used the Schottenuhr design. The movement sometimes had an alarm, and different dials were used, including the wood shield type, porcelain, metal, carved, bronze frame, and others. In the mid-19th century an eight-day Schottenuhr was introduced.[6]

THE CLOCK DEALER AND TRAVELERS

The sales of the clocks by clock peddlers started as early as the clockmaking in the Black Forest. In the beginning the sales were handled by the glass dealers, who also handled wooden utensils and straw hats that were made in the Black Forest. Those dealers all travelled by foot through the countryside, often working between twelve and fourteen hours a day.

The clocks sold very well and made a profit for the glass carrier that was between four hundred and five hundred percent of their original cost. The clockmakers wanted some of that profit and decided to do their own travelling. In the winter months, when the snow was deep in the Black Forest, the clockmaker and his whole family concentrated on producing clocks. In the summer months he went on the road by himself.

The History of Black Forest Clocks 23

Side view of the early 1760 cuckoo clock. Wood plate movement with wood wheels, wood arbors and some steel pinion gears. Rope driven with the early weights and primitive cuckoo bird. Maker unknown. (Antique Clocks Publishing Archive.)

Side view of the wood wheel cuckoo with painted lacquered dial (Lackschild). Strikes on the hour on a metal bell with cuckoo call. Metal escapement wheel. Maker unknown. Circa 1780. (Deutsches Uhrenmuseum, Furtwangen/Schwarzwald, photo Callwey, München, Germany.)

24 Black Forest Clocks

The Potato Eater, or "Knödelfresser," made about 1880. Enamel dial. Carved, painted figure at top opens mouth and moves arm as if eating. (Photograph from slide bought at Deutsches Uhrenmuseum, Furtwangen.)

Picture frame cuckoo clock with painting on metal, hunter motif. Enamel dial. Circa 1870. (Photograph from slide bought at Deutsches Uhrenmuseum, Furtwangen.)

The clock peddler, or "Uhrenträger," made about 1860. Painted hollow tin figure with wood base. The clock he is carrying on the front has a small brass plate and gear movement that actually keeps time and will run for one day. The clock on his back is only the shell of a shield clock. (Photograph from slide bought at Deutsches Uhrenmuseum, Furtwangen.)

Number 558 J.B. Beha shelf clock with 6 tune music box in the base of the clock. Bone hands and numbers. The case is inlaid with decorative carving stained in black. A small carved eagle sits on top. Number 161 is also written in pencil on the backboard, side doors, inside of the case on both sides of the cuckoo doors, and also on the lead backing of the pendulum bob. Repair dates of Feb 1884 and June 9, 1885 are written in pencil on the backboard. The music box tunes will each play for about 35 seconds after the cuckoo sounds on the hour call, except at 1 o'clock. The finials are not original.

The wood plate, eight-day, double fusee movement with 6 tune music box.

26 Black Forest Clocks

Monk shield animation clock made about 1780 by Matthäus Hummel. The painting depicts the monastery in which the monk lives. Figure moves when the clock strikes on a metal bell on the hour. Wood plate and gear, movement with brass. (Deutsches Uhrenmuseum, Furtwangen/Schwarzwald, photo, Callwey, Munchen, Germany.)

An early Black Forest wood plate movement with brass gears and wood arbors that was used for a shield clock with bell strike, and would run for one day. Circa 1830-40.

Two side views of the movement showing the wood arbors.

Travel at that time was not easy. Roads were rough and conditions were hazardous. It was not long before the clockmakers grew tired of this extensive travelling. Besides it was better to have the master at home maintaining production. Many clockmakers turned to another family member or close associate to do the travelling. This associate, of course, needed to be able to make adjustments to the clock after it was hung at the customer's home. This is how the foundation was laid for the trade of the clock peddlers.

The clock peddler (carrier) wore a special uniform: dark leather pants, a long dark jacket with no collar and a single row of buttons, wool stockings, and boots. Under the jacket was a red vest. A scarf was worn around the neck. The hat was a large round black farmer's hat. On his back he carried clocks without their shields (disassembled). On a belt over his shoulder and down the front was one or more totally assembled clocks to show as samples. His food sack, money sack, I.D. card, and a big umbrella were tied to his backpack. He also had a walking stick (cane). To travel by foot twelve to fourteen hours a day with such a load was quite an accomplishment.

In 1720, Jakob Winterhalter, from Gütenbach became the first clock peddler. Jakob purchased his wares from Anton Ganter, a clockmaker in Neukirch. He travelled as far as the Rhine region and Holland and combined the clock dealings with the peddling of parrots. He was known as the "clock-and-bird dealer." Other prominent early dealers were Thomas Bärmann of Schallach, Joseph Kammerer of Furtwangen, and Joseph Ganter of Neukirch.

Part of the payments from the customers included such things as a meal and a room for the night. The overnight stay was very popular because the customer could test the clock overnight to make sure it operated correctly before the dealer went on his way. The practice became so traditional that even a very wealthy clock dealer would spend the night or have a meal. It seemed that no place was too far for them to travel.

Eventually, it seemed too much to return to the Black Forest every time they sold all of their clocks. This is when they arranged to have "clock packers" (wholesalers) in various towns, saving extensive travel time. This created more job opportunities for the people of the Black Forest. It became very popular to be a clock peddler. Later, many of them worked together as unions and shared their profits and losses equally. They travelled extensively throughout Europe.[7]

28 Black Forest Clocks

A shelf cuckoo and monk clock with an 8-day, triple fusee, wood plate movement by Johann Baptist Beha.

The shingled roof, spiral finials, and gothic top.

The monk bell ringer located at the base of the clock.

The label on the backboard translated:
J.B. Beha and Sons
Cuckoo Clock Fabrication
Eisenbach
(near Neustadt)
Black Forest

The cuckoo doors, bird, and top carving.

The wood plate movement viewed through the side door.

A side view of the clock.

The triple fusee, wood plate, brass gear movement.

FROM SMALL CLOCKMAKER TO FACTORY:
The American System

Not taking into account individual farmers, who initially carried on clockmaking only on the side, little had changed economically between 1700 and 1850 in the Black Forest clock business.

The entire clock production was handled in small family businesses, each consisting of only a few skilled workers. This is referred to as "in-house trade" or "in-house industry." Businesses of suppliers of parts and semi-finished products, including clock faces and case manufacturers as well as clock face painters, were organized in the same fashion.

After the mid-19th century, technological progress penetrated all areas and changed the existent economic and social structure. The biggest changes in clockmaking in the Black Forest occurred in the middle of the changeover from making clocks in homes to making them in the factory buildings. Around 1840, in the northeast United States around Connecticut and Massachusetts, a clock industry was booming. Eli Terry, Chauncy Jerome, Seth Thomas and other lesser known clockmakers were producing mainly longcase clocks patterned after the English styles. The problem for the Black Forest clock industry was that the Americans had begun to mass produce these clocks, and were offering strong competition in the world market, taking business away from the Black Forest area.

In response to the transition to factory-type clock manufacturing in the United States, occurring earlier and giving them a strong and threatening competitive position over the Black Forest clockmakers, the clockmaker school in Furtwangen (founded in 1850) tried to promote the greatest possible division of labor, the use of machines, standardized clock work types, and rational manufacturing methods.

Around 1860 the Black Forest clockmakers began producing clocks made after the American system, and the clocks were made as cheaper exports.

Ogee style, cottage, and school clocks were copied by the Black Forest clockmakers. The school clocks were made in the Eisenbach region with quality movements. The German clocks were lined on the inside with blue paper and the backboard was removable.

At first, Black Forest clockmakers disapproved of these cheaply made American-style clocks, but later found out they had to make the same clocks to survive in the world market.

But the anticipated effect failed to come about and was only realized by a few independent master clockmakers. As soon as they had acquired the necessary capital, they turned to more rational manufacturing methods. The clockmakers school dissolved in 1863. It was a devastating blow for the clockmaking and case manufacturing trades. The woodcarving occupation had no artistic input after that. In Hornburg, Triberg, Furtwangen, Vöhrenbach, and Waldkirch, a few small businesses went into mass fabrication and tried to stabilize the down-sliding market.

The in-house industry gradually lost its traditional character through the increasing use of machine labor and it began to develop into large scale enterprises. These businesses usually employed less than 300 people, however the Badische Uhrenfabrik factory was an exception, hiring more than 700 workers.

Independent clockmakers joined these clock factories. The factories increasingly dominated and displaced the in-house industry, which quickly fell behind the time. Many previously independent master clockmakers became homeworkers for the clock industry, assembling clocks or manufacturing parts.

Two different groups were formed among the in-house industrialists. One group consisted of small independent masters--clockmakers, parts makers, clock face painters, woodcarvers, and case joiners who handled purchasing and sales for their own accounts. The other group consisted of homeworkers, who produced parts or assembled clocks, and who were dependent on clock manufacturers. They received the material, or rather, the parts, from their clients, and their pay was based on the number of pieces delivered.

The small clockmakers, who often specialized in custom-made clocks, had a particularly hard time since they had to compete directly with the large scale enterprises. Precision cuckoo, "Jockele" clocks, and dual alarms figured most importantly in this respect, as they often had a special clientele.

In 1861 Erhardt Junghans and his brother Xavier started making American styles of clocks in Schramberg. By 1864, they were mass producing these clocks and the rest of the Black Forest clockmakers followed suit. Phillip Haas & Sons started this mass produced American style of clock in 1867. These clocks eventually were exported to the United States, with instructions written in English. Some of the Black Forest clock companies even changed their names to English names.

Phillip Haas used Teutonia Clock Manufactory in St. Georgen. Others were the Union Clock Company and Badenia Black Forest Clock Manufactory in Furtwangen.[8]

The History of Black Forest Clocks 31

An early Black Forest clock movement, wood plate, brass gear, wood arbors and rope driven, time-and-strike on a gong. Circa 1820.

Two photos showing the side view of the brass gear, wood arbor movement.

The wood plate, brass gear, steel arbor, 8-day, double fusee movement with gong and Beha label. (Photo by Doug Barr, Berea, OH.)

An early Johann Baptist Beha double fusee, eight-day, wood plate movement shelf clock with cuckoo and calendar. Overall height is 18 inches tall. The case is inlaid and has an enamel dial. (Photo by Doug Barr, Berea, OH.)

The label, Johann Baptist Beha in Eisenbach near Neustadt, in the Black Forest. (Photo by Doug Barr, Berea, OH.)

The History of Black Forest Clocks 33

The 30-hour Beha wood plate, brass gear movement. (Photo by George R. Edwards)

Number 302 wall cuckoo made by the Beha factory. This clock was originally purchased by Edwin Jones (born December 13, 1849, died March 28, 1932). He was the House Steward at Ashburnham Castle, England for 55 years and he purchased the clock while on holiday in Switzerland in the late 1800s. After his death, all personal items, including the clock, were willed to George L. Reese (the grandfather of George R. Edwards) in Philadelphia, PA, and in the 1950s was given to the mother and father of George R. Edwards. He remembers seeing it on the wall when he was a young child. His mother gave it to him in 1987. This clock is a nice original example of a Beha clock passed through the same family. (Photo by George R. Edwards)

The backboard of the clock marked model number 302. (Photo by George R. Edwards)

A side view of the movement with one train of gears.

A simple time-only wood plate, brass gear 30-hour movement. Circa 1850.

THE WOODCARVING SCHOOL

The knowledge related to this once very important craft school is almost gone today. The carver's occupation is among the oldest of the woodworking trades. Since older times, most household and other articles for everyday use on the Black Forest farms were made of wood. The wood carving trade was also closely related to the clockmaking industry.

According to the art taste of the time, around 1850 the box clocks which were lavishly decorated with carvings and cut wood became very popular. During this time, the popularity of the regulator clocks with turned and carved adornments was very strong. Only a few of the very top masters were able to fulfill these specific needs, because the simple wood cutter was not able to master this demanding technique. In order to fulfill the growing demand, a lot of suppliers from other regions had to be engaged.

Up to the year 1863, the Baden clockmakers school was able to supply these needs. Not only did the school work on the technical development of clock works, they also addressed the tasteful embellishments of the clocks, the fabrication of the housing, and the wood carving.

The Viennese World Fair in 1873 proved that the clockmakers trade could not compete if the education of the artistic carver was not being continued. In 1877 the Baden government reopened the clockmakers school and a carvers school at the same time. The grand opening of this institution was on July 1, 1877 in the trade hall in which the clockmakers school was located. The carvers school had the assignment to educate first class carvers with special emphasis for the Black Forest clock

A time-and-strike wood plate, brass gear, 30-hour movement with the strike on a wire gong. Circa 1850.

Side view of the movement.

Side view of the movement.

36 Black Forest Clocks

Early shelf cuckoo with oil painting on metal, enamel dial, and double fusee, 8-day wood plate movement. Case wood is burled. Circa 1860. Johann Baptist Beha.

The eight-day, wood plate, double fusee movement.

A picture frame cuckoo with oil painting on metal. Wood plate, 30-hour movement. Circa 1870.

The History of Black Forest Clocks 37

A group of 30-hour, wood plate, blinking eye clocks with oil painting on metal. All three are circa 1865-70.

38 Black Forest Clocks

An example of a wood plate, brass gear, 30-hour movement used with a shield type face with bell strike. Circa 1850.

Side view of the movement.

industry and to support the industry of the Black Forest with their expertise.

At first the curriculum included the following mandatory classes: free hand drawing, model building in plaster, wax, and clay, and wood carving. In 1898 a specialty of furniture and cabinet building was added. The principal of the first carver's school was the wood sculptor Professor Johann Koch, who received his artistic education in Munich and was known as a very qualified expert.

Under the guidance of Professor Koch, from 1877-1906 the carver's school in Furtwangen developed into a highly acclaimed training and guidance institute for the Black Forest arts and crafts movement. In the small school approximately 40 students would be educated at a time.

The carver's school also trained students in the manufacture of other things such as boxes, chests, and figurines. These stylish and tasteful samples were available in a plentiful and constant display for use by craftsman and manufacturers. With this arrangement, it was hoped that a flat mass production would be alleviated and the carvers would produce first quality workmanship. Especially high workmanship was produced in the cabinetmaker department, with furniture richly decorated with carved adornments in a variety of styles. In the year 1882-1883 the carvers school received its own school house. The most fruitful time for the carving school was around 1891.

The yearly report of 1904-1905 stated that during the preceding year, as in previous years, the instruction at the carving school had strictly followed its curriculum, which regulated instruction in carving, modelling, and drawing for all apprentices, with the exception of the carpenter apprentices—which was the only way to instruct the apprentices who usually came inexperienced in the proper understanding of form and technique.

The carpenter apprentice workshop had a special, well-designed course that was strictly followed. It provided the apprentices with a proper foundation so that they could be drawn upon early for manufacturing entire pieces of furniture.

The enormous change resulting from the introduction of the modern style led woodcarving and the arts and crafts as a whole, to new avenues. The carving school began to emphasize this new style to a much greater extent than it had in previous years, taking pains to meet the call of the new times.

The History of Black Forest Clocks 39

Shield cuckoo clock. The door for the cuckoo was made part of the painting, and when the door is closed, one hardly notices it. Circa 1850. The hands are later. (Gerd Bender, *Die Uhrmacher des hohen Schwarzwaldes und ihre Werke, Band I*, Verlag Müller, Heimatmuseum, Schwenningen.)

Painted dial shield clock, time-and-strike, and alarm, 30-hour wood plate, brass gear, steel pinion movement. Circa 1860.

40 Black Forest Clocks

Unusual cuckoo and quail clock in that the birds are exposed at the top of the clock. Roses, birds and leaf motif. Eight-day, triple fusee, brass movement. Case is 24 inches tall. Circa 1880. (Photo by Doug Barr, Berea, OH.)

The triple fusee movement with new bellows. (Photo by Doug Barr, Berea, OH.)

Black Forest animation clock with the figure of a King drinking from a bottle and glass at the top of a barrel. 30-hour brass movement, while the animation train must be wound about every three hours. The King takes a drink about every five minutes. Circa 1875. (Sundial Farm, Greenlawn, NY.)

Flute clock made by E. Wehrle. Eight wooden pipes for this model made about 1880. Bone hands would have been a more intricate carved style. (Photo by Doug Barr, Berea, OH.)

Picture frame clock made in the Black Forest about 1870. Porcelain dial with a thin brass frame around it. Wood plate movement with brass gears and steel arbors.

Picture frame cuckoo with oil painting on metal and enamel dial. The cuckoo door was missing on this clock and the replacement has not yet been repainted to match with the rest of the painting. 30-hour wood plate movement. Circa 1860.

The History of Black Forest Clocks 43

Picture frame clock with reverse glass painting, which tends to flake off as the years pass. Circa 1880. (Photographed at a NAWCC convention, San Jose, CA.)

Picture frame clock with reverse glass painting, with dog and hunter scene. Wood plate movement. Circa 1870.

During the long winter evenings, apprentices, demonstrating assiduousness and zeal, voluntarily worked until 9:30 p.m. under their instructors' supervision. They were allowed to feel free to consult the school library and to do their homework assignments at the school. This was an offer that should certainly not be underestimated, considering that it meant these indigent apprentices could use the heated school rooms as a second home.

In response to an invitation by the Hornberg trade board and with the permission of the grand-ducal trade school board in Karlsruhe, in summer 1904 the school sent a large number of different works to the trade exhibition in Hornberg, to demonstrate the achievements of the carving school to larger industrial circles in the Black Forest. The wood carvings they exhibited there included clock faces, frames, consoles, mirrors, panels, as well as several small pieces of furniture in the modern style. They earned high praise from the public. The many woodcarvers working for carving shops in Hornberg, in particular, expressed interest in these works made in the modern style.

During the school year the school board took the opportunity to visit the industrialists in the Black Forest. Furthermore, the school board received sufficient funding from the grand-ducal trade school chamber for visiting woodcarving schools in Berne, Brienz, and Meiringen, as well as the arts-and-crafts museums in Berne and Zurich.

The carving school kept a large number of carvings for its own collections to use as teaching aids.

At the end of the school year, the school board provided the graduates with suitable placement in different carving shops and carpentry shops so these young people could, in addition to earning a good income, have the opportunity to further their vocational skills.

A Singing Bird clock in a nice architectural beechwood case with silvered and gilded frets applied to the center of the 18 cm dial and front of the case. The winding holes of the eight-day movement are concealed by shutters. The movement strikes the hour on a wire gong and the bird sings at each hour and moves in three directions. Height: 29''; width: 14½''; depth: 9½'' The clock was made by Emilian Wehrle. (Clocks of Distinction, Sydney, Australia).

The writing on the case translates: "With the speed of an arrow, it has flown away, eternally fixed in the past. Slowly comes the future."

Lenzkirch wall clock with brass decoration and ½ hour strike movement. The case is walnut and measures 45 inches in length. Circa 1880. Fifty-four pieces of brass applied to the case.

Lenzkirch mantel clock. Case is mahogany and measures 19 inches tall. Circa 1880.

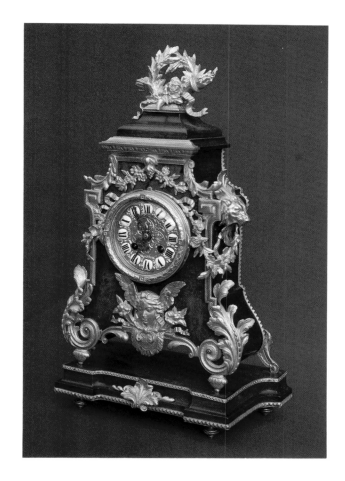

Very large wall clock made by the Lenzkirch factory about 1890, supposedly made for a jeweler in Prague, Czechoslovakia. The clock was for many years part of the Greensboro, North Carolina clock museum. The case is mahogany with brass ornamentation including female figures as pillars on the door, and it measures 53 inches tall by 32 inches wide.

Lenzkirch mantel clock measures 22½ inches tall, made from oak with 14-day movement.

Kasten-uhr, or box clock, with reverse painted dial and painted case. 30-hour wood plate, brass gear movement with hour and half hour strike. The weights are new.

Picture frame clock with reverse glass painting. Mountain and church scene at the bottom. Circa 1880. (Photographed at a NAWCC convention, San Jose, CA.)

Curriculum

Instruction at the school was comprised of five subjects:

1. Woodcarving
2. Modelling in clay and wax
3. Freehand drawing and drafting, etc.
4. Shaping from models
5. Carpentry

The summer weekly schedule included:
Carving—44¾ hours
Modelling, including shaping from the models (two sections)—18¾ hours
Drawing—12 hours
Carpentry—54 hours

The winter weekly schedule was as follows:
Carving—40¾ hours
Modelling, including shaping from the models (two sections)—16¾ hours
Drawing—12 hours
Carpentry—49 hours

Apprentices who had not yet successfully completed a trade school were obligated to attend the systematic instruction in the trades in addition to their carving studies. The exceptions to this were freehand drawing, drafting, and modelling, since these subjects were given at the carving school.

As a rule, apprentices began school on May 1. Every youth, who was mentally and physically healthy and who had reached the age of 14 and had graduated from primary school, could be admitted to the carving school.

At the turn of the century woodcarving and modelling were taught by the extraordinarily capable instructors Philemon Rombach and Josef Münzer.

After the decease of Prof. Johann Koch in 1907, director Eugen Hauffe, who had previously taught drawing, took over as headmaster of the school and served from 1907 to 1931. The school moved to a new location in 1926.

From 1932 to 1935 the director of the clockmaker school, Emil Jäger, simultaneously managed the administration of the carving school.

A nice, open well, freeswinger style made by the Adolf Hummel Regulator factory. Freiburg in Breisgau. The oak case measures 42 x 18 inches. Time-and-strike, eight-day movement. Circa 1890.

Picture frame clock with wood plate, brass gear movement that runs for one day, driven by weights. Embossed style around the dial. Circa 1870.

48 Black Forest Clocks

The oil painting on wood which depicts the maiden in the garden with the trumpeter.

The side view of the Säckingen trumpeter clock showing lower side doors and hole for winding the music box located in the base of the clock.

See page 223 for other great views of this musical 9 horn trumpeter clock made by Emilian Wehrle, Furtwangen, about 1890.

The side view of the Säckingen trumpeter clock showing the movement side door.

An unmarked alarm clock in the shape of a castle. Circa 1880.

After 1935, the carving school was under the provisional supervision of the director of the trade school, Friedrich Hub, and in 1938 the carving school was closed and the entire property sold. If one will take the time and study closely some of the carved and architectural styles of clocks which are shown in this book, one might agree that this carving school had a definite impact on the techniques of the Black Forest artisan and some very beautiful examples were produced and are still in existence today.[9]

Despite the state's initiative of re-introducing a clockmaker school as well as founding a wood-carving school in Furtwangen, the decline of the in-house industry could not be halted. There were approximately 1,400 independent masters around 1873. By 1882, the number decreased to 1,034. In 1888 the government tried to preserve the in-house industry. An attempt was made to establish a cooperative of the individual branches of the in-house trade, however the trade boards would not take to the idea of a cooperative, and with a feeling of resignation they took the view that the decline of the in-house industry could not be stopped and accepted the reality of the large scale industry.

In 1905 there were only 132 independent masters left.[10]

A nice, unmarked mantel clock. The case is made from oak and measures 17 inches tall. Silvered chapter ring with one piece of beveled glass in the door. Eight-day movement. Circa 1910.

Footnotes

[1] E. John Tyler, *Black Forest Clocks*, pp 3-7, N.A.G. Press, Ltd. London, 1977.

[2] Gerd Bender, *Die Uhrmacher des hohen Schwarzwaldes und ihre Werke*, pp. 240-264, Verlag Müller, Villingen, 1979 revised.

[3] Gerd Bender, *Die Uhrmacher des hohen Schwarzwaldes und ihre Werke*, pp. 185-192, Verlag Müller, Villingen, 1979, revised.

[4] E. John Tyler, *Black Forest Clocks*, p. 7, N.A.G. Press Ltd., London, 1977.

[5] Dr. Wilhelm Schneider, *Die eiserne Kuckuckuhr*, pp. 37-44, "Uhren", Verlag Callwey, München, 1989.

[6] E. John Tyler, *Black Forest Clocks*, pp. 3-32, N.A.G. Press Ltd., London, 1977.

[7] Gerd Bender, *Die Uhrmacher des hohen Schwarzwaldes und ihre Werke Bd.II*, pp. 261-276, Verlag Müller, Villingen, 1978.

[8] Gerd Bender, *Die Uhrmacher des hohen Schwarzwaldes und ihre Werke*, pp. 113-121, Verlag Müller, Villingen, 1978.

[9] Gerd Bender, *Die Uhrmacher des hohen Schwarzwaldes und ihre Werke*, Bd.II, pp. 97-110, Verlag Müller, Villingen, 1978.

[10] Gerd Bender, *Die Uhrmacher des hohen Schwarzwaldes und ihre Werke*, pp. 111-113, Verlag Müller, Villingen, 1978.

An unmarked German shelf clock with original bracket. Brass decoration on the case which measures 24 x 16 inches. Bracket is 12 inches in length. Circa 1895.

An artist conception of a clockmaker working in a Black Forest cottage workshop. Wood etching from Wilhelm Hahn about 1850. (Antique Clocks Publishing Archive.)

The movement on the left is a 30-hour time-and-strike, spring driven movement. On the right is an 8-day time and alarm. Both are wood plate. The alarm is removed from the right movement. Counting the barrel as a gear, the escape wheel makes four gears on the left movement, and makes for five gears on the right movement. The 8-day is also larger. (photo by Barry Malkemes)

German drop dial style clock. Walnut veneer is stained to match the mahogany surround of the wood dial. Eight-day, wood plate, double fusee movement. Clock is 20½ inches tall. Circa 1870. (Photo by Barry Malkemes)

German gallery clock made in the Black Forest. Painted dial. No mark on the movement. (Photo by Barry Malkemes)

CHAPTER 2

The Cuckoo Clock

Cuckoo clocks were developed by Black Forest inhabitants because they desired another occupation, besides farming, that could bring in much needed funds during the long winters of this region. The cuckoo clock eventually was marketed worldwide and became very popular as a novelty clock and did bring in more revenue for the Black Forest farmers. The first cuckoo clock made in the Black Forest has been credited to Franz Anton Ketterer of Schonwald (born 1671) in the year 1730. While the idea of the cuckoo clock may have earlier origins in another region, it became the most closely clock identified with the Black Forest region.

The cuckoo clocks were very primitive looking at first. Many were made with square wooden dials painted with water colors, or with the painted arch wooden dials. They had wood plate and gear movements. As the years went by, cuckoo clocks became more decorative in appearance. Some were made in picture frame style cases with oil paintings on the front, and the cuckoo door above the dial, in the painting. The cuckoo bird was automated and some were feathered. The beaks and wings of the bird were automated. The themes of the paintings varied from hunting motif to family scenes, etc. The imaginations of the painters were unlimited and many themes were depicted. Biedermeier style cuckoo cases were also made with porcelain columns and enamel dials on the front of the clock cases. Architectural and carved cases were mass produced before the turn of the 19th century.

The most famous cuckoo clockmakers were Johann Baptist Beha of Eisenbach, Fidel Hepting of Gutenbach, and Theodor Ketterer (1815-1884) of Furtwangen. But the clockmakers only completed the clock; many others contributed to it. He received parts from five or so people and put the clock together. Birds were carved and painted mainly by women. There were bellow pipe maker, the movement makers, wood carvers for the cases, and finishers.

It is difficult to estimate the exact age of Black Forest cuckoo clocks. Many of the clocks were unsigned, with no label or movement stamp. However, many clocks in this book are dated. The age was determined by the case style, the style technology, and materials used in the movement, the finish of the case, and other clues.

The light colored Black Forest spruce wood was used for the backboard of the cuckoo clock because it increased the sound level of the clock's spiral gong. Beech and linden woods were used for wood carving because of their fine grain, and these woods were very suitable for carving. Walnut was also used in later carved cases, but was more expensive and harder to carve. Beechwood was also used for the wood plates of the earlier cuckoo movements.

Earlier cuckoo clocks have a cuckoo call that is very similar to the call that the bird actually makes. The call is generated by two pipes tuned to A and F. The bellows that create the wind are located at the top of the pipes. During the release of the strike train, the bellow top is moved up and down by a wire actuated from the lifter wheel. The weight of the bellows create wind when moving back to the normal position, and the wind moving through the pipes creates the cuckoo call. The shorter pipe sounds first, followed by the larger pipe.[1]

Around 1884 the most important product of the Black Forest was the small carved cuckoo clocks, some with music boxes. In the small villages in and around Triberg, it is known that an estimated 13,500 men and women were engaged in the making of clocks. This area was then known, and now, as clock country. There were the home clockmakers in their large houses as well as the

54 Black Forest Clocks

factories where the gears, pinions, and movements were assembled into the cases made by the carpenters. The clock assembly method was such that the clock passed from one person to another, one person putting in the winding barrel, then another the train of gears, another the escapement, and another person then fitting the movement to the case. Finally the clock was packed for sale. In the large houses or cottages, there could have been many families engaged in the clockmaking business, with each family occupying one or two rooms of the cottage. There were perhaps ten to twelve families in one cottage. The people used their various specialties when making the clocks. There was the wood cutter preparing the wood for the cases, the case maker, the shield maker, the painter, the gear maker, chain maker, spring maker, the movement maker, dial maker, the wood carver, and the people who finalized the case.

From Germany, Black Forest clocks were sent all over the world. England bought many trumpeter clocks, cuckoo clocks, and regulators. Other clocks were sent to Belgium, Holland, Russia, Spain, Portugal, etc. The U.S.A. bought many trumpeter clocks, cuckoo clocks, and musical clocks.

There were clock schools in the Black Forest area. In June, 1877, two years of clockmaking experience was required for admission to the school in the Gewerbe-Halle. The course lasted one year and the students studied arithmetic, algebra, geometry, physics, mechanics, technical clockmaking, drawing, bookkeeping, etc. They also had to complete thirty hours of practical clock work per week in the clock shop. This was done in the afternoons. The school was small and classes were held in two small rooms. At this time (1877) eighteen pupils attended, most being admitted free or supported by exhibitions given by neighboring towns in the Black Forest.[2]

The cuckoo clocks pictured in this section are today highly collectible and are fast disappearing into collections all over the world.

Larger cuckoo clock with a four tune music box that plays on the hour. The case is 60 inches long, carved from linden and walnut wood. The bird at the top is an Auerhahn, or German grouse. Two deer are at the bottom of the case which was made as a wedding present for T. Leifeld in the Black Forest about 1890. He was a vegetarian and did not like the dead animal hunter motif seen on many clocks, but rather wanted the live animal scene on his clock. Bone hands with porcelain numbers on the dial. This is one of the most exquisitely carved larger examples made by the Black Forest artisans.

The carved musician that appears at the hour when the clock plays one of the four tunes on the music box.

The dial with porcelain numerals and bone hands.

The Auerhahn at the top of the clock.

The two deer at the bottom of the case.

56 Black Forest Clocks

Kastenuhr, or box style, cuckoo clock with porcelain columns. The painting on the case has been retouched. Thirty-hour wood plate, brass gear movement. Case is 21 inches tall by 10 inches wide. (Cleveland 1986 NAWCC National Convention.)

Carved wall cuckoo with one-day cast brass movement, unsigned. Bone hands. Walnut case. Circa 1900.

Wall cuckoo clock with thirty-hour cast brass movement. Case is 30 inches tall by 20 inches wide. The front shield carving is made from one piece of wood. Unusual motif. Circa 1900.

The Cuckoo Clock 57

Larger cuckoo clock. Case is 32 inches tall by 19 inches wide. Pheasant and pine tree motif. Thirty-hour cast brass movement. Circa 1880. Movement is spring driven. Bone hands. Case made from linden wood. (Cleveland 1986 NAWCC National Convention.)

Small cuckoo clock, made about 1880. Rooster at the top of the clock. Movement is wood plate with brass gears and arbors. Photographed at a NAWCC Convention, San Jose, CA. Bone hands and numbers. Original pendulum.

Small wall hanging cuckoo with delicate leaf and vine carving. Wood plate, brass gear movement. Bone hands. Circa 1880. Original pendulum. Smaller, simpler made bone numerals.

58   Black Forest Clocks

Small thirty-hour cuckoo clock with brass movement, and eagle and deer motif. Bone hands. Cuckoos at the hour and once at the half hour. Circa 1890.

Cuckoo and quail clock made about 1900. The carving is nice but a simpler style of wood-carving. Spring driven movement runs about thirty hours. Bone hands.

Small thirty-hour cuckoo clock with bird and leaf motif. Enamel numerals. Circa 1890. The top and shield are carved from linden wood.

The Cuckoo Clock 59

An earlier polished case with brass inlay shelf cuckoo. Circa 1860. Eight-day, double fusee, wood plate, brass gear movement.

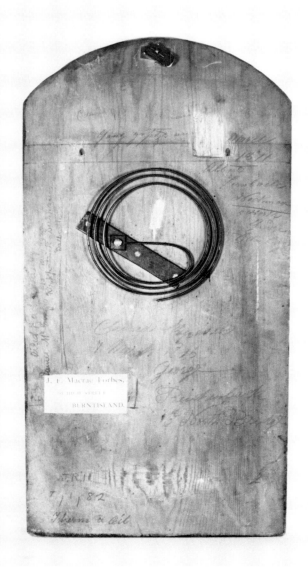

The backboard and gong and the many repair signatures and labels that tell the history of this clock.

Cuckoo clock with the fox and grapes and eagle motif. Nicely carved but the movement has some stamped gears which indicate a later clock. Circa 1915. Hands made of celluloid.

The backside of an early eight-day, wood plate, fusee movement. This movement was not made by the Beha factory. Note the difference in the springs that hold the stop gear rachet in place.

Cuckoo and quail clock with a bird, pear, and grapes motif. About 24 inches in length. Bone hands. Maple leaf design, thirty-hour movement. Circa 1900.

A closeup of the stop gear and rachet. The setup on the right is original while the wire on the left has been changed.

Cuckoo clock with eagle at the top and deer at the bottom. Nicely carved. Circa 1900. Cast brass movement.

A wall cuckoo clock with hunters motif, nicely carved with thirty-hour cast brass weight driven movement with some stamped gears. Circa 1915. The antlers on the deer head are original.

Cuckoo clock made about 1900. Oak leaf design with three deer head and one rabbit and bird at the bottom. Cast thirty-hour brass movement. Nice, intricate carving.

Carved wall hanging cuckoo clock with deer at the top, made about 1900. Thirty-hour cast brass movement. Bone hands. No maker is found anywhere on the movement or case.

Small wall cuckoo with thirty-hour wood plate movement. Vine and leaf motif. Circa 1870.

Front of the wood plate movement, which is unmarked.

The backside of the movement showing the count wheel, strike hammer and wire linkage to activate the cuckoo bellows.

Side view of the wood plate movement with recovered bellows and wood bird.

Cuckoo and quail clock with one door at the top. Cuckoos on the hour and the quail calls on the half hour. Has a shift mechanism to activate cuckoo and quail from one train. Wood plate, brass gear movement which runs for thirty hours. Twenty-two inches high. Exquisitely carved bone hands. Cuckoo doors are not original.

64 Black Forest Clocks

Cuckoo and quail clock made about 1915. Thirty-hour brass movement. The carving is much simpler in this time period with only seven large leaves, although the birds are carved more in detail than would be expected. Bone hands.

The bird on the side of the case.

The bird at the top of the clock.

Wall hanging cuckoo clock, case made from oak and carved with very few leaves, but more intricate birds. Circa 1900. Cast brass thirty-hour movement. Hands are not original.

Nicely carved cuckoo clock made about 1910 and thirty-hour brass movement driven by two weights, one for the time side and the other for the cuckoo, which strikes with a gong on the hour and half hour. The bone minute hand is not original.

Larger wall cuckoo carved from linden wood with eagle and birds' nest motif. Case is about 33 inches tall. Circa 1900. Bone hands. The eagle at the top is very intricately carved.

The carved eagle at the top of the case.

The young birds feeding at the bottom of the case.

Another spring driven wall cuckoo with eagle and flag motif. Wood plate, brass gear movement. Case is 22 inches tall. Circa 1870. The oak leaf is predominant in the shield of this clock. (Cleveland, 1986 NAWCC National Convention)

Smaller cuckoo clock with leaf, rose, and bird motif. Bone hands. Circa 1910. The case is carved from linden wood and is finished with a dark stain.

Large cuckoo clock with the hunter motif. Length of clock is about 48 inches. Original throughout. Circa 1900. Eight-day cast brass movement. Antlers are original.

68 Black Forest Clocks

Spring driven wall cuckoo with eagle and hunters motif. Porcelain numbers with bone hands. Eight-day movement. Circa 1900.

The spring driven eight-day movement with newer style bellow. AF is stamped on the movement.

The Cuckoo Clock 69

Larger, weight driven, thirty-hour cuckoo with music. Cast brass movement. Bone hands. Eagle and hunters motif. Circa 1890. A very nicely carved original clock.

The live eagle motif.

Musical cuckoo clock. The cuckoo appears every 15 minutes at the top and dancers appear at the bottom door at the ½ hour and hour. Spring driven, cast brass movement. Case is 31 inches tall by 13 inches wide, and is made of oak with brass decoration. Etched metal dial.

Black Forest Clocks

Cuckoo clock with thirty-hour cast brass movement. Carved with deer and the farmhouse motif. Bone hands. Circa 1895.

Later shelf cuckoo clock made about 1900 with an eight-day movement with brass plates and brass gears. The time or gong train of the movement is spring driven while the strike and cuckoo train are fusee driven. The carving on this case is a little more simplified, but still well done. The two birds below protecting their nest of eggs. The bellows are new or replaced. Bone hands with celluloid numerals. The case carving is walnut.

The brass spring driven and fusee movement. The cuckoo bird is mounted on the wire with a small metal bracket. The earlier cuckoo birds are mounted by a wood platform and screw. This is a good way to estimate the age of a clock. The wood mount types are earlier.

Shelf cuckoo with St. Bernard dog motif. Spring and fusee brass movement will run eight days. Case is 29 inches high. Circa 1900. The minute hand is not original. Bone grommets are inserted in the winding holes. Many times they are missing because someone was too enthusiastic in winding the clock and they were easily broken.

Shelf cuckoo clock with eagle and deer motif. Eight-day brass fusee movement. Case is 31 inches high by 21 inches wide. A very nicely carved clock. A more simplified style of bone hands was introduced as time passed.

The brass fusee and spring movement which runs for eight days. Bellow tops and pendulum are not original.

Shelf cuckoo clock with rose, daisy, and leaf motif, with thirty-hour cast brass movement. 21 inches high by 14 inches wide. Circa 1900. The shield carving became simpler as the years passed and the exquisite wood carving capabilities were lost with the passing of time. (Cleveland, 1986 NAWCC National Convention)

Shelf cuckoo clock with a thirty-hour, brass, spring driven movement. Case is 21 inches high by 14 inches wide. Door is not original. The carving is of a simpler style. The bird at the top is not as intricately carved as others that are pictured in this book. Circa 1900. (Cleveland, 1986 NAWCC National Convention)

Shelf cuckoo clock with the vine leaf and fox motif. Circa 1880. Thirty-hour movement. The way the shield has been finely carved with intricate vines indicates an earlier style or better skilled wood carver. The leaf at the top and the cuckoo door are not original, however.

The Cuckoo Clock 73

The dial with the carved bone numbers and hands.

Shelf cuckoo clock with St. Bernard carved at the top. Eight-day, brass plate fusee movement. Case is 29 inches tall. Kamerer and Kuss label on the backboard. Circa 1885. A good example of intricate bone numerals and hands.

Shelf cuckoo clock with the fox and bird motif, a thirty-hour case brass movement made about 1880. Nicely carved bone hands.

Smaller shelf cuckoo, about 15 inches high with thirty-hour cast brass movement. Simplified carving indicates a later clock, made about 1900. Bone hands are not original.

Cuckoo with music box in a castle style case with brass decoration and a spring driven movement. Circa 1880. Silvered metal dial with metal hands.

Shelf cuckoo clock with carved butterfly at the top. Simpler carving indicates perhaps a later clock. Thirty-hour cast brass movement. Circa 1900.

Shelf cuckoo clock with the pheasant motif. Eight-day, brass fusee movement. Case is 29 inches high by 25 inches wide. Circa 1880. The clock is very intricately and nicely carved by a talented wood carver. Bone hands.

Wall cuckoo with weight driven thirty-hour, cast brass movement. Bird, leaf, and rose motif. Circa 1910.

Thirty-hour spring driven wall cuckoo with three-bird motif. Circa 1900. This carving is nice, however it is simplified compared to the earlier wood plate carved cuckoos made by the Beha factory about 1870.

Cuckoo with bear motif on the shield and a bear used for the pendulum bob. Circa 1910.

Hunter motif with eagle. Cuckoo with thirty-hour, weight driven, cast brass movement. Circa 1910.

A later cuckoo, the case in the shape of a chalet with two figures that are painted. Circa 1920.

A nicely carved cuckoo with bird at the top and nest of birds at the bottom. Thirty-hour, spring driven, cast brass movement. Circa 1900.

Another nicely carved cuckoo with weight driven, thirty-hour, cast brass movement and two birds with leaf and berry motif. Circa 1900.

78 Black Forest Clocks

Carved shelf cuckoo with deer and fern motif. Case measures about 24 inches tall and is nicely carved. Full relief carving. Circa 1890.

The bellow pipes and spring driven, cast brass movement.

Bahnhaüsle cuckoo and trumpeter clock. Cast brass movement, spring driven. Circa 1900.

The Cuckoo Clock 79

A simple wall cuckoo and quail clock with bird at the top and leaf motif made from linden wood. Circa 1910.

Shelf cuckoo in a fretwork style motif with spring driven fifty-hour movement. Unmarked. Circa 1875.

Cuckoo and quail clock in a Bahnhaüsle case style. Wood plate, brass gear, thirty-hour movement. Circa 1870.

Closeup detail of a swan carving. (Barry Malkemes photo.)

Carving from a thirty-hour, wood plate cuckoo. Top shield and pendulum bob match. Linden wood. (Barry Malkemes photo.)

Carving of a thirty-hour, wood plate cuckoo. Carving of walnut. (Barry Malkemes photo.)

Thirty-hour shelf cuckoo. Cast brass movement with no makers' name. Carved leaves on case. Overall height is 18 inches.

Larger hunters motif cuckoo and quail clock with stag and two dogs. The case is stained very dark and measures about 38 inches. Wreath around the dial. Circa 1900.

Wood plate with brass gear, one-day movement, wall cuckoo. Leaf vine, and berry motif. Intricate bone hands. Mountain goat at the top of the clock. A very pretty and intricate style of wood carving. Circa 1870.

Wall cuckoo clock with deer head, rabbit, pheasant and dog motif. The shield is unusual in that there is no top roof gable separate from the bottom shield. It is all one piece with the animals attached to it. Circa 1900.

A simpler style wall cuckoo with very little carving on the shield and crow at the top roof gable. Cast brass one-day movement. Circa 1910.

A small wall cuckoo with wood plate movement and blue porcelain numerals on the dial. Circa 1870.

Fretwork style shelf cuckoo with fifty-hour, spring driven, brass plated movement. Again, all carving is original on this clock. Circa 1900.

Elaborately carved wall cuckoo with weight driven, cast brass movement, which runs for one day. Circa 1900.

Small shelf cuckoo with fifty-hour, wood plate movement and vine leaf motif. Bone hands, and the case is made from oak. Circa 1870. The top gable would probably have had a few vines going from leaf to leaf, however they are broken off and have not yet been repaired. Case is about 17 inches tall.

Shelf cuckoo; the double fusee, wood plate, brass gear movement removed and being restored at the time of this photo. Beechwood case. The clock was probably made by a factory other than the Beha factory, as the movement is different.

Shelf cuckoo that measures about 28 inches tall. A different style of hunters motif with a boar at the bottom of the case. The minute bone hand is broken and the top of the clock would have had a little more carving, but it was broken away at some point in time.

Weight driven wall cuckoo with eagle at the top and lion at the bottom, made about 1900.

A Gothic style shelf cuckoo, fifty-hour, brass plate movement that is not marked or stamped with any factory identification. Circa 1900.

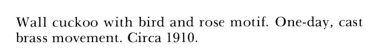

Wall cuckoo with bird and rose motif. One-day, cast brass movement. Circa 1910.

86 Black Forest Clocks

Gothic wall cuckoo made from oak with wood plate, weight driven movement that runs for one day. Bone hands. The finials and top crest are original. Circa 1870.

Spring driven shelf cuckoo, first sold August 2, 1881. The label on the back also "Charles E. Pelhum, Goldspring, New York," a jeweler. The front carving is linden wood. Thirty-hour movement, wood plate, brass gears.

Backside of the wood plate, spring driven movement and original bellows. (Antique Clocks Publishing Archive).

CUCKOO AND QUAIL BIRDS

The cuckoo bird actually measures about 13 inches. Most have a slim body and beaks that are small and bent. Their wings are full and sharp edged, and are black with white and black spots. They have a rounded tail and their legs are short and covered with feathers. The male usually has ash grey feathers on his back, and his chest is striped. The female's neck is red. The eyes, beak, and legs are yellow. Cuckoos are found in Africa, Asia, and northern Europe. They have excellent flying endurance. The cuckoo bird never builds its own nest, but settles near the nests of other smaller birds who hatch the cuckoo eggs. The cuckoo is then raised by the unsuspecting smaller bird. The male cuckoo always guards his territory with the consistent cuckoo call.

The quail bird was used with the cuckoo clocks for the quarter hour strike. The quail is a native of Europe and Asia, and is closely related to the American quail.[3]

An earlier carved wall cuckoo clock with fifty-hour wood plate, brass gear movement. Circa 1870.

The cuckoo bird being fed by its adoptive parent. (Antique Clocks Publishing Archive).

The cuckoo bird which measures about 13 inches in length. (Antique Clocks Publishing Archive).

The quail which was used in the cuckoo and quail clocks. (Antique Clocks Publishing Archive).

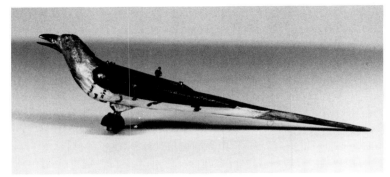

An example of an earlier cuckoo bird carved from wood and painted. This style of bird could be found on cuckoo clocks around 1865-1880 and mostly seen on wood plate, brass gear movements. This particular bird is from a clock made by Johann Baptist Beha from Eisenbach.

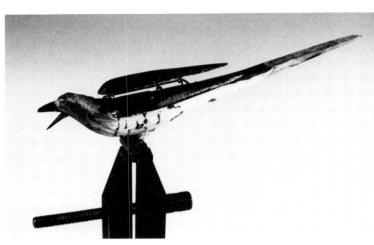

View of the same bird with wings and beak in action. Note the wood platform that the bird sits on, which indicates it went with an earlier clock.

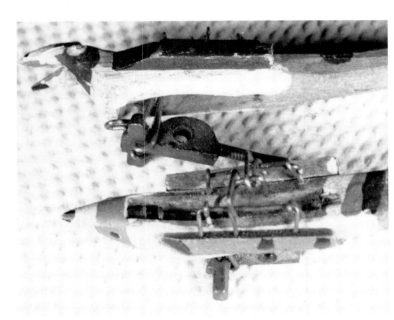

Cast brass mounts of turn-of-the-century birds. (Photo by Barry Malkemes)

The Cuckoo Clock 89

An example of a later cuckoo bird taken from a clock made about 1910-20. Note the metal bracket below and the inner wing wire linkage showing, which indicates this bird was made for a later clock. The bird is also much thicker in appearance, and is not as delicately carved.

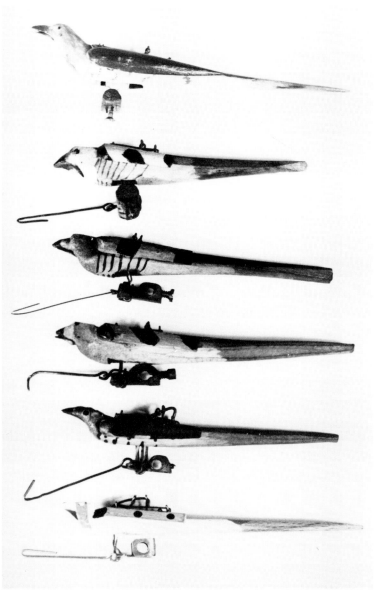

A group of six wooden cuckoo birds, the early bird at the top (circa 1870) and the one on the bottom from much later (circa 1940). You can see the progression from the early, sleeker, more delicately carved bird at the top to the simpler bird with squared wings at the bottom.

Close-up of bird linkage. (Photo by Barry Malkemes).

90 Black Forest Clocks

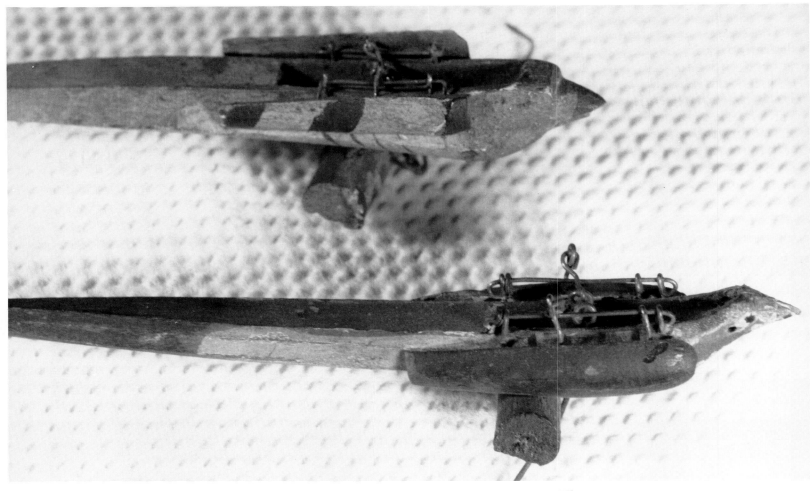

Two birds of early vintage. Note the wood mounts. The bird at the bottom is earlier and has nicer detail and finer wire linkage. (Photo by Barry Malkemes)

Photo of a World War I vintage bird. The mount is cast brass and the body is made of tin as are the tongue and wings. The underside of wings are painted in a polychrome red. (Collection of Bill Heron, photo by Barry Malkemes)

HANDS

Hands were made in a very simple style in the beginning and, as time went on, they became fancier with the clock styles. At first, simple, carved hands made of wood were used on the wood wheel clocks. In the mid-18th century poured brass hands were made and used on the shield clocks (Lackschilduhren). Brass hands, cut from sheets of brass, were also used on shield clocks, flute clocks, calendar clocks, and baroque shield clocks. Iron or metal hands, painted black/blue, were used on the picture frame clocks, Jockele clocks, and mantel and floor-standing clocks. Carved cuckoo clocks had hands carved from bone, and ivory was used for special clocks such as the trumpeter. Pewter hands have also been seen on trumpeter clocks.[4] Celluloit hands were used on later cuckoo clocks.

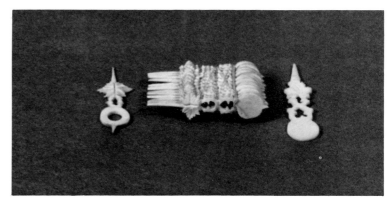

Bone hands in the middle, still tied with the string for the clockmaker to use on the new clocks that were being finished. All are the same pattern.

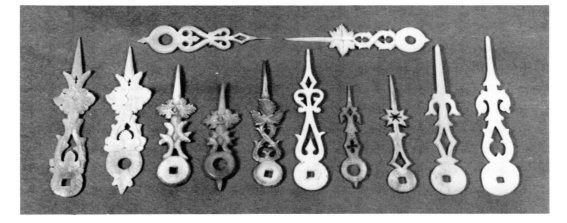

Various patterns of bone hands, the first five on the left bottom row could be found on earlier clocks, and the later simpler pattern of the five on the right are mainly found on later Cuckoos.

WEIGHTS AND PENDULUMS

The weights for Black Forest clocks were used to drive the time and strike sides of the movement. The Black Forest clockmaker used many different styles of weights through the years, since their appearance changed with the period of furniture. At the beginning they were used only to run the clock, however as time passed they became ornaments and added to the style of the clock. The clockmakers, through the years, used the following for weights:

1. Rocks that were taken from streams.
2. Cannon balls.
3. Metal weights.
4. Hollow glass weights, filled with sand.
5. Hollow wood weights, filled with sand.
6. Iron weights.
7. Brass weights.
8. Pine cone weights made of metal.

Many weights were lost over the years and there is confusion about which weights go with which clocks. The early Black Forest clocks had rocks as weights. The clockmaker tried to polish these weights to make them look more attractive; they used the rocks taken from rivers or streams as they already had a polished look to them. They seemed to have the proper shape for these early clocks and were tied by a rope, or a net-like wire was built around the rock and attached. In most cases, a hook was used to hold the weight. They sometimes used small cannon balls for the weights, secured with wire and metal.

When the clockmaker started to make chiming movements that required two weights, they realized they needed the long shape of a weight so that they would not interfere with each other and stop the clock. These were then made of metal or lead and were round or squared. They were also made out of green colored bottle glass, by local glassblowers. The cowtail clock needed exact weight to run it accurately as it was very delicate, so they used hollow wood or glass weights and also made them from colored clay. These materials were easy to regulate for the exact size and the weight. They were simply filled with the proper amount of sand to do the job. The empty weight shells were also easier for the clock peddler to carry!

At the end of the 18th century the shield clocks (Lackschilduhr) were fitted with weights made mainly of iron, melted into shape, which was round at the bottom and narrow at the top of the weight. The hook was added to the weight at the time of the melting, and they sometimes left open space in the center of the weight so they could later regulate its mass by pouring more lead into it. Sometimes the dealer who bought clocks from the clockmaker bought the clocks without weights and would make them himself or had them made in the town in which he lived to save on transportation costs.

Later, lead weights were used. Very fine brass was wrapped around the lead for appearance, leaving a very fine seam. Brass covered lead weights without the seam are of a later style. The brass covered weights were very popular with the picture frame clocks (Rahmenuhr). Dark picture frames with brass or gold trim were made more beautiful by brass weights and pendulum.

The most famous weight is the pine cone style used for a variety of clocks, but mostly on the carved cuckoo wall clocks. This pine cone style is still used on today's mass produced cuckoo clocks.

The scale of the pine cone was not as detailed on the older weights. The patterns are better detailed on the newer weights. Before 1870 the weights were very basic without a lot of decoration, and after this year (1870) the Black Forest clockmaker matched the weights more with the decor of the clock. These weights were also made in other parts of Germany.[5]

Some of the pendulums used for cuckoo and other Black Forest clocks, such as picture frame and two-note trumpeter clocks, are also shown in this chapter.

Various cone weights used for cuckoo clocks.

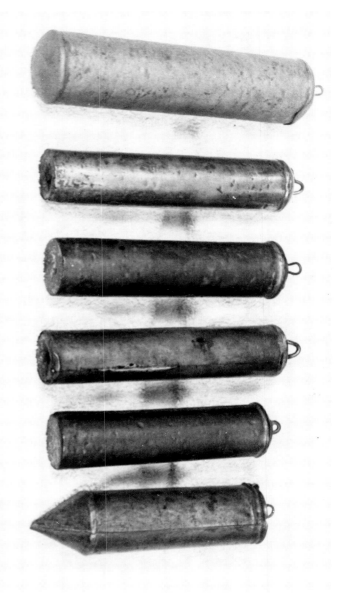

Various brass weights filled with lead. They were used on picture frame clocks. The ones with the seam in the brass are early.

The Cuckoo Clock

Weights generally used for cuckoo clocks.

Other weights used for picture frame and cuckoo clocks. The third and fourth weight could have been used for a cuckoo clock.

Weights used for cuckoo clocks.

Other pendulums used for cuckoo clocks. The one on the right is generally found on the Bahnhäuse style case.

94 Black Forest Clocks

Various brass pendulum bobs used for picture frame and cuckoo clocks. The fifth from the left could be used on a three-note trumpeter clock also. The second from the left on a picture frame clock.

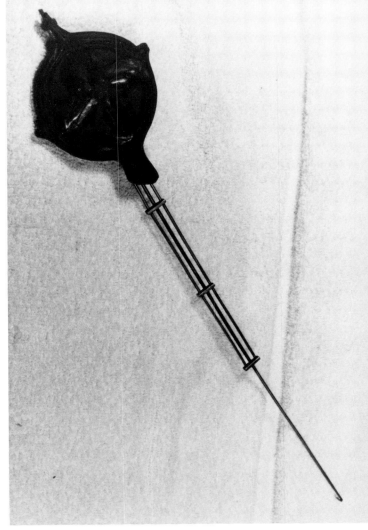

A pendulum used for a wood plate, thirty-hour cuckoo movement.

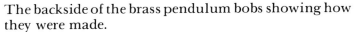

The backside of the brass pendulum bobs showing how they were made.

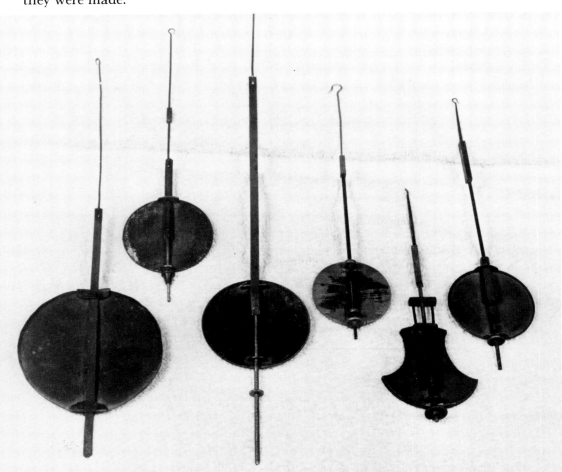

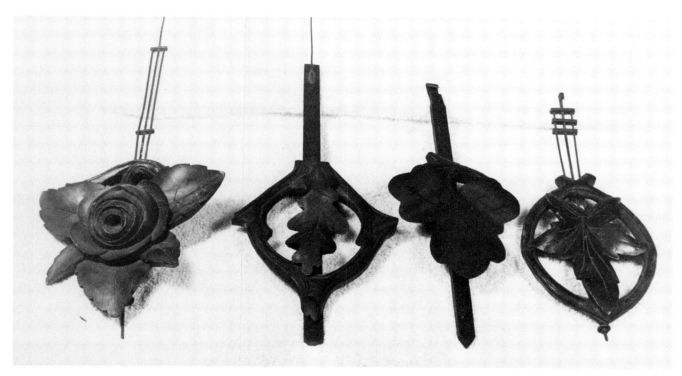

Four pendulums used for brass plate cuckoo movements.

WOOD AND BRASS PLATE CUCKOO MOVEMENTS

The wood plate, wood wheel cuckoo movement dates from about 1730. The wood plate, brass gear cuckoo movements appeared around 1840 and the Lyra type cast brass movement dates from about 1865. The stamped plate and gear cuckoo movement appeared around 1900.[6]

Footnotes
[1] Karl Kochmann, *Black Forest Clockmaker and the Cuckoo Clock*, pp. 3-42, 99, Antique Clocks Publishing, Concord, 1987.
[2] *Horological Data Bank*, The Watch and Clock Museum of the NAWCC, Inc.
[3] Karl Kochmann, *Black Forest Clockmaker and the Cuckoo Clock*, pp. 13-16, Antique Clocks Publishing, Concord, 1987.
[4] Gerd Bender, *Die Uhrmacher des hohen Schwarzwaldes und ihre Werke*, pp. 437-442, Verlag Müller, Villingen, 1979 revised
[5] Gerd Bender, *Die Uhrmacher des hohen Schwarzwaldes und ihre Werke*, pp. 430-437, Verlag Müller, Villingen, 1979 revised.
[6] Karl Kochmann, *Hamburg American Clock Company*, pp. 2-22, Antique Clocks Publishing, Concord, 1980.

Eight-day, wooden plate double fusee, brass gear, steel arbor cuckoo movement, made by Johann B. Beha. Circa 1875-1880. However, the movement is unmarked.

96 Black Forest Clocks

The backside of the Beha movement with count wheel, stop gears and original cuckoo pipes and bellows.

A left side view of the movement.

A right side view of the movement.

The Cuckoo Clock

The front side of a wood plate, spring driven, thirty-hour cuckoo movement with brass gears and steel arbors. No maker's signature or stamp is found on this movement.

The backside of the movement showing the count wheel, stop gears, and old original bellows. Note that the bird sits on a wood platform which is earlier.

A side view of the same movement.

The backsides of two earlier cuckoo movements with the plates held together by pins. All gears are cast. The straight bar type plate movements were found from 1865 through about 1900.

Stamped "Sears Roebuck & Co., Germany." Thirty-hour cast brass lyre movement. (Photo by Barry Malkemes)

An example of the backside of a later cast brass cuckoo movement with some cast gears and also some stamped gears. The nuts that hold the plates together indicate a later cuckoo movement, as does the cuckoo bird which is attached above by a metal bracket versus a wood platform. Also more of the wire linkage shows above the cuckoo's wings. A mixture of cast and stamped gears indicate a later clock.

Three earlier cast brass cuckoo movements. The left one is found frequently (Lyre). The squared plate movements were also found frequently.

The Cuckoo Clock 99

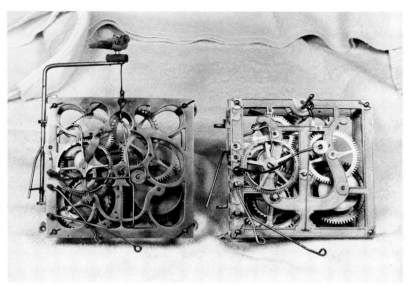

On the left, an unusual cuckoo movement with skeletonized plates. On the right, an unusual etched pattern on the plates of the cuckoo movement.

A cuckoo and quail movement on the left that can be disassembled in sections (unusual) and a later stamped gear cuckoo movement on the right held together by nuts instead of pins.

Three different size cuckoo movements with cast brass plates, but some gears are stamped. The nuts to hold the plates together indicate a later movement.

CHAPTER 3

Johann Baptist Beha of Eisenbach

Johann Baptist Beha (1815-1898) was born in Oberbränd (Black Forest), the son of Vinzenz Beha (1784-1868). Vinzenz Beha was a talented cuckoo clock maker, producing the shield clock type cuckoos (Lackshild). It appears that the area around Oberbränd was very popular in relation to cuckoo clock makers. Around 1810 there were others besides Vinzenz Beha: Conrad Ganter, Josef Rombach, Johann Wildi, and from neighboring Eisenbach there was Martin Beha.

The shields for the early cuckoos were produced by the Kirner family from Kleineisenbach, the Stegerer and Reich families from Eisenbach, and Frans Mellert from Bubenbach.

The expert maker of the wooden cuckoo birds was Josef Schwarz from Oberbränd. The cuckoo pipes were made by the clockmakers themselves; the wood for these pipes and some of the wood for the cases was supplied by the Brugger family from Unterlenzkirch.

In 1845 Johann Baptist Beha set up his clock shop in Eisenbach. He had been making cuckoo clocks on his own in his father's workshop. His father, Vinzenz, was also his instructor. Vinzenz was well known for the special quality of the clocks that he produced. Johann inherited the house in which he had set up his shop.

It is uncertain if Johann Beha learned everything about clockmaking from his father or if he studied under other clockmakers. He also had a close relationship with Rupert Maurer of Eisenbach and Peter Wehrle in Dittishausen.

Rupert Maurer was an expert in making clocks patterned after English clocks and probably had earlier worked with Winterhalder and Hofmeier on eight-day clocks. He had made a clock in the 1850s with strike on the quarter hour on bells. Rupert was awarded a silver medal at an exhibition in Villingen in 1858. He also displayed several complicated eight-day table clocks in München in 1854 for which he received much praise.

A Black Forest shelf clock, circa 1845, designed from the Vienna mantel clocks with eight-day wood plate movement. Alabaster columns and a small oil painting at the top of the clock, also a small window at the bottom of the clock. Johann Beha also made shelf cuckoos in this same exact manner with 50-hour and eight-day wood plate, brass gear movements. (Dr. Wilhelm Schneider, Regendorf, Germany)

Rupert and his partner Felix Höfler of Eisenbach started their own clock factory in Eisenbach called Maurer and Höfler. Rupert and Johann Beha probably exchanged many ideas, especially regarding the English influence on their clocks.

Peter Wehrle was already an excellent case maker and in 1850 delivered his cases to both Johann Beha and Rupert Maurer. Wehrle may have received his education on making cases in Wien (Vienna), Austria.

Bahnhaüsle cuckoo with colored lithograph which depicts a yarn spinner motif, made by Johann B. Beha. Circa 1865. (Dr. Wilhelm Schneider, Regendorf, Germany)

Earlier shelf cuckoo with inlay made by Johann Beha about 1860. Eight-day, wood plate, brass gear, fusee movement. Case is 21 inches tall, and highly polished. Sold by Camerer Kuss & Co. in London, England. (Cleveland, 1986 NAWCC National Convention)

Early inlaid shelf cuckoo which stands 19 inches tall, made by Johann Baptist Beha about 1860. The inlay is made of brass, mother-of-pearl, and turquoise inlay. (Cleveland, 1986 NAWCC National Convention)

Black Forest wall clock with oil painting on metal. 20 inches long by 15 inches wide with enamel dial. Wood plate, brass gear, eight-day movement. Circa 1860. Made by the Beha factory in Eisenbach. (Cleveland, 1986 NAWCC National Convention)

Cuckoo with inlay, an enamel dial, and an oil painting on metal on the front of the clock, made by Johann Beha. The movement is wood plate with brass gears and steel arbors, and will run for one day. Circa 1860. (Cleveland, 1986 NAWCC National Convention)

Johann, it seems, had contact with other firms such as Lorenz Bob, who also made cuckoo clocks, and Matthä Sättele of Eisenbach who was a member of a commission of experts regarding clocks made with English construction at the clockmaker school in Furtwangen.

Johann's contact with all of the aforementioned people was probably the key to the great success that later came to the Beha clock company. Johann also developed a close business relationship with Gordian Hettich of Furtwangen. He was a good salesman and a very good source of information between the clockmakers. His personal connections around 1845 worked out well for him, even though the economic and political times were not good. At that time in the Black Forest, the economy suffered from a low period in agriculture brought on by poor crops. The Civil War during 1848 and 1849 complicated matters more. There are no records to document, but from the period 1845-1850 Johann Beha seemed to survive and keep his company going. From the records that could be found, 1848 seemed to be less profitable.

With his own funds, Johann Beha made 365 cuckoo clocks between 1839 and 1845, their worth totalling approximately 2,000 Gulden. These were made in his father's workshop, where cuckoo clocks were mainly produced. Most of the cuckoo clocks made used different technology and design than the ones made previously; very few were similar. From 1849-1842 shield cuckoo clocks were produced mainly with some movements made with wooden wheels or half wooden and rope driven. These were very popular at this time. Beha clocks were sent to Spain, Glascow, Scotland, Madrid, and other countries. Many clocks were also sold to Gasthauses (restaurants or inns).

Between the period of 1842-1845 some of the clock cases appeared to be designed by Peter Wehrle. The Kastenkuckucsuhr (box case cuckoo clock) was very popular. It usually had a roof top (triangular) and was made with alabaster columns and mirrors on the front of the case. This was a totally new type of cuckoo clock for this time period; it was also very expensive for this time. These clocks had pewter dials that were made and painted by Johann Hogg in Eisenbach. With these clocks the cuckoo bird was moving or mechanical, but had no wings.

In March, 1844 there were other innovations for Johann Beha—a fifty-hour spring driven cuckoo and a weight driven cuckoo in a wall model. In May, 1844 a table clock with alabaster columns and a fifty-hour spring driven movement was made. This clock was very similar to the Viennese table clocks, with a window at the bottom of the case.

Johann Baptist Beha of Eisenbach 103

J.B. Beha wall thirty-hour cuckoo. 23½ inches high. Mahogany veneer. Enamel dial. Gilt frame behind glass. 1850. (Photo by Barry Malkemes)

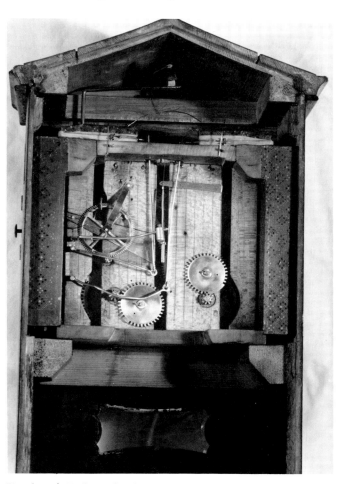

Back of Beha clock. Note the winding stops. The hammer is incorrect. Notice bird mounting, which is early. (Photo by Barry Malkemes)

Close up of the stop gears, which are missing on many clocks. (Photo by Barry Malkemes)

The front side of the movement showing the spring caps. (Photo by Barry Malkemes)

On the backboard of the case, the deteriorated Beha label. (Photo by Barry Malkemes)

104 Black Forest Clocks

A number 512 Johann Baptist Beha shelf cuckoo with eight-day double fusee movement and a six-tune music box in the base. The beechwood case measures 20 inches tall with vine and leaf motif with an intricately carved bird at the top. Bone hands with celluloid numbers. One of the six tunes will play for about 20 seconds after each hour's cuckoo is sounded (except for 1 o'clock). The music box is believed to have been made by Ducommun-Girod. Frederic-Guillaume William Ducommun first made music boxes at about the age of 25 in 1820. He had settled in Geneva around 1815. In 1835 a directory describes him as "a maker of all kinds of music boxes" under the firm name of Ducommun-Girod living at Tour-de-Boel 62. His father, David, and his brother, Henry, worked for him. Frederic died on April 8, 1862, but his two sons, Louis and Jean, ran the business until 1868. They won a bronze medal at the Paris Exposition of 1867. Ducommun & Cie was formed in May, 1869, when Louis Ducommun joined into partnership with Mittendorf, Louis. This lasted five years at the corner of rue Kleburg and rue du Mont Blanc, near Nicole Freres. Louis died around 1890. Jean continued with a workshop in Geneva and made musical box parts until about 1885. He died in December of 1899. The music boxes made by Ducommun-Girod were of very high quality and always play in a very pleasant manner. Usually one or more of the tunes on these music boxes had sustained trills in the upper register of the musical comb.

The Ducommun-Girod label printed in black on which the musical tunes were written. The words at the bottom mean "dampers of steel" and "easy to wind."

The eight-day Beha movement with six tune music box.

The Ducommun-Girod music box removed from the clock.

This clock is again a vine leaf and berry motif with a differently shaped shield applied to the case, made by Johann Beha. It is 27 inches tall. The Beha factory made many different styles of cuckoo clocks. They did not follow the other factories who mass produced clocks with cheaper carvings and movements. Rather they stayed with the concept that the better clocks would last longer, and sold to the people who could afford them in those days. This clock was made about 1875.

A picture frame cuckoo with porcelain dial made by Johann Beha. The cuckoo door is not original. Circa 1860. (Cleveland, 1986 NAWCC National Convention.)

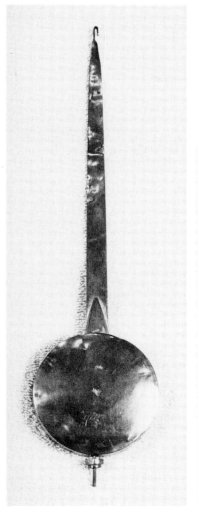

The pendulum used for the Beha shelf clock, made of brass with lead filling in the round disk.
The back of the pendulum showing the lead filling.

106 Black Forest Clocks

Gothic style shelf cuckoo with three figures that strike bells for the chime. Triple fusee movement, wood plate brass gear, steel arbors, made by Johann Baptist Beha. Bone hands. Circa 1870. (Cleveland, 1986 NAWCC National Convention)

The triple fusee, wood plate movement with three figures and cuckoo bird. The bellow tops have been replaced.

The carved figures that strike the bells of the Beha clock.

That Johann Baptist Beha was the originator of the spring driven cuckoo clock, is questionable. It appears not, as there is a spring driven table cuckoo clock in the Geymullerscholss, Wien Museum which was made in the second half of the 18th century and has a moving cuckoo bird and two cuckoo pipes. However, he was the first in the Black Forest to produce them. The earlier cuckoo clock was probably made on an individual basis, was extremely rare, and did not contribute to further cuckoo development. Johann Beha did successfully produce them, and was responsible for their popularity.

Johann made mantel or shelf cuckoo clocks that had English influence in their case design. They had unusually large movements and were similar to the early English bracket clocks. The cases had no carving but were veneered and highly polished.

It is not known exactly when Johann Baptist Beha made the first eight-day cuckoo, but it was probably between 1845 and 1850. It was hard to tell the difference between Johann Beha clocks and Theodore Ketterer clocks, since the same casemaker worked for both clockmakers. The later clocks were not signed with any name or signature, which also made it difficult to identify the clock maker. It should be said here that Theodore and Anton Ketterer clocks never gained the same respect that the Beha cuckoo clocks achieved. The Beha's were extremely popular and famous. Nevertheless, Ketterer did make very high quality clocks, even though the factory was very small.

About 1850 Johann Beha started producing wall cuckoo clocks in square cases, some with and some without moving eyes. The fronts of the clocks were painted with various scenes and designs on metal, and were made with spring driven movements around June 1850. Beha put the cuckoo door in the painting, above the dial.

The aforementioned clock is not in the same category as the later tin plate front Bahnhäusle cuckoos that were made by Beha from about 1860. These were made mainly for Russia and the Eastern countries. It does not appear that Johann Beha was also the first to use this type of art on a clock in the Black Forest.

Polished, cased, mantel clocks with cuckoo and double fusee, wood plate, and brass year movements became very popular. Some of the cases were inlaid with brass, pewter, mother-of-pearl, turquoise, etc.; some had enamel dials, others painted metal dials. These were very popular around 1860.

The carved cuckoo clocks began to appear about 1865 and became extremely popular. Whether Johann Beha was the first to produce these carved clocks is not certain. They may have first been produced by Theodore Ketterer of Furtwangen around 1861. Beha certainly developed many styles that sold well and became very popular as the years went by.

A wide assortment of carved cuckoo clocks made by Johann Beha was exhibited at the Centennial in Philadelphia in 1876. Beha received a gold medal award at the international exhibition in Vienna on August 18, 1873. It appears he was the only cuckoo clock maker who was honored in this way. He also received honors at international exhibitions in London (1862), Paris (1867), and at regional exhibitions in Villingen (1858) and Karsruhe (1861). His firm (sons included) received a meritory medal by the U.S. Centennial Commission for the clocks shown in Philadelphia. He and his sons, Lorenz and Engelbert, further received honors at exhibitions in Karlsruhe (1877), London (1885), Freiburg (1887), Chicago (1893), and Strasbourg (1895).

Wall cuckoo clock with brass decoration and similar to model 399, which was made by the Beha factory. Case is 24 inches tall. Bone hands. Circa 1875. "Baahnhäusle" style. An architectural style case is not unusual, however carved cases seemed more popular.

108 Black Forest Clocks

Blinking eye cuckoo clock made by Johann Baptist Beha about 1880. The eyes of the man in the metal painting, the wolf, and the dog move with the swing of the pendulum. Double fusee, wood plate, brass gear movement. Case is stained very dark, almost black. Intricately carved bone hands.

The dial with bone hands and the blinking eye of the wolf and dog.

The double fusee movement. The pendulum and bellows are not original. The wire apparatus above and below the movement are connected to the escapement at the top and the pendulum at the bottom to achieve the blinking eye movement.

Beha's major accomplishments include the following: He was the first to build the Schotten cuckoo clock movement in a clock case. He was the first to make Black Forest wall and shelf cuckoo clocks with spring driven movements of the fifty-hour type, and between 1845-50 the first double fusee, eight-day, wood plate, brass gear movements were first made by him.

Beha also made oil painted, picture frame cuckoo clocks where the cuckoo door was not part of the picture, but rather part of the roof gable, above the painting. During 1850-1870, until the beginning of the Franco-Prussian War, cuckoo clock styles increased along with the production.

A little should be said about the exhibition in 1876 at the Philadelphia Centennial. Mr. G.S. Lovell was the agent of the Black Forest clock manufacturers and he had imported clocks from the Beha factory since 1873. On March 6, 1876 forty-six cuckoo clocks were sent to Mr. Lovell, with an invoice of about $636. The lowest wholesale price was $4.77, the highest was $32.60. The retail prices at the exhibition were higher and the Beha clocks sold for between $15 and $60. Of the clocks sent to the exhibition, thirty were wall types and sixteen were shelf models and all but two had wood plate, brass gear movements. Six were cuckoo and quail or quarter strike. One was a monk clock ringing the angelus on coil gongs, and another was a double fusee movement with moving eye mechanism. Clocks resembling models 312, 509, 512, 539, 610 and 671 pictured in this chapter were all part of the exhibition.

All of the clocks exhibited at the Centennial Exhibition in Philadelphia were sold. Before this exhibition, heavily carved cuckoo clocks with eight-day fusee movements were rarely sold in the U.S.A. The Bahnhausle style model #124 was a smaller clock and seemed to sell very well in the U.S.A. through 1880. Models 539 and 576 were the top selling styles. Number 361, which is pictured in this chapter, seemed to sell throughout the years. Large clocks were slow to sell, as they were too expensive to ship to the U.S.

After the exhibition in Philadelphia, Lorenz and Engelbert Beha joined their father's firm as partners and it was renamed Johann B. Beha and Söhne. The Behas did not believe in the mass production of cuckoo clocks as did other larger factories such as Junghans, Kienzle, Phillip Haas & Söhne, Werner, and some of the smaller factories such as Dold, Hilser, Gordian Hettich Sohn, etc. These firms developed their cuckoo clock production in response to the demand for an inexpensive clock with inexpensive carvings. The Beha clocks were not made in this way, and Johann Beha did not intend to start this method of

Thirty-hour spring driven wall cuckoo clock made about 1870. Case is about 15 inches long with original pendulum. Wood plate, brass gear movement. Clock was made by the Johann B. Beha factory in Eisenbach. There was a small cross at the top of the case surrounded by a leaf design, but it is missing.

production after so many years of producing quality clocks. Other factories that stayed with top-of-the-line production clocks for the upper class were Emilian Wehrle & Co., Winterhalder & Hofmeier, Maurer & Höfler, Lenzkirch, and L. Furtwängler & Söhne (LFS), all located in the Badenia Black Forest. The man who built the quality cases for Beha, E. Wehrle and LFS was Augustin Tritscheller from Furtwangen.

At the exhibition in Chicago in 1893 the following German or Black Forest companies exhibited clocks: Johann B. Beha & Sohne; L. Furtwangler & Söhne (LFS), Thomas Haller, Maurer & Höfler, Jahresuhrenfabrik Schatz, Gebrüder Junghans, Friedrich Mauthe, August Schwer, Emilian Wehrle, C. Werner, Gebrüder Wilde, and Winterhalder & Hofmeier. Beha sent only eight clocks to Chicago, two were carved and

110 Black Forest Clocks

Beha number 690 wall cuckoo with cross on the top and very small intricate leaf and vine pattern. Thirty-hour wood plate, brass gear, spring driven movement. Bone hands. The case measures 13½ inches. There should be another leaf at the lower right but it was broken off and lost. This style of carving is very delicate.

The nicely carved pendulum for the Beha clock.

The back of the pendulum with the heavy round metal disc attached to the wood carving.

the rest were the neo-renaissance or architectural style. Beha sold all of the clocks exhibited in Chicago, but this apparently did not help future sales in the U.S. as only the inexpensive carved cuckoo clocks were exported in the future. The architectural style of cuckoo clock was not very successful in the U.S. Beha did, however, receive an award for excellence of quality, variety of styles, and fine decoration at the Chicago exhibition. Sales through Mr. Lovell continued until 1898, though he purchased only 472 clocks during that five year period. Johann Baptist Beha died in 1898 and his sons, Lorenz and Engelbert, continued with the making of cuckoo clocks. It should be said that Johann Baptist Beha was considered *the* cuckoo clock maker of the Black Forest, and he was described in many trade papers as very respected and having a talent similar to that of Lorenz Bob. His clocks are much sought after today by Black Forest clock collectors.

A good way to identify Beha clocks is to inspect the movement. The double fusee wood plate usually had the same stop gear levers and springs, and the model number was normally written on the backboard. The carved cases had a certain look to them; ones copied by other factories were not of the same quality as a Beha clock. If one studies the movements, cases, carvings, etc., one should be able to identify Beha clocks more easily. The firm seldom used a label to depict the factory name.[1]

It should be stated that in this chapter almost all of the written research on the Beha factory has been performed by Dr. Wilhelm Schneider of Regendorf, Germany. It has taken many years to accumulate all of the history of this important cuckoo clock factory and he should be commended for his efforts.

Footnotes
[1] Dr. Wilhelm Schneider, Monika Schneider, *Black Forest Cuckoo Clocks at the Exhibitions in Philadelphia 1876 and Chicago 1893*, pp. 116-131, NAWCC Bulletin, Vol. 30 #253, Columbia, PA, 1988; and Dr. Wilhelm Schneider, *Frühe Kuckucksuhren von Johann Baptist Beha aus Eisenbach im Hochschwarzwald*, pp. 45-53, "Alte Uhren", München, 1987.

Gothic number 733 shelf cuckoo by Johann Baptist Beha in an unrestored condition. A small piece of carving is missing at the top and the spiral finials have numerous tips that are broken off and missing. Double fusee, eight-day, wood plate movement. Circa 1880.

Wall cuckoo clock with thirty-hour wood plate movement. Split cuckoo door. Bone hands. Clock probably made by the Beha factory. 19 inches tall. Circa 1880. A nice original example with the exception being the numerals.

Shelf cuckoo clock made by the Beha factory. Case number 539. With its eagle, deer and dog motif, the case is 24 inches tall by 17 inches wide. This clock sold at the exhibition in Philadelphia in 1876 for $17.16. The hands and numerals are not original.

Gothic style shelf cuckoo with spring drive movement, wood plate, probably made by the Beha clock factory. The hands are not original. The numerals are made from bone. Circa 1870.

Johann Baptist Beha of Eisenbach 113

Cuckoo and quail clock made by the Beha clock factory about 1875. Thirty-hour, wood plate, brass gear movement. Ivory hands. The case number is written on the back, but is illegible. Similar to Beha number 671.

Beha shelf cuckoo number 512 sold on March 24, 1882 by W.T. (this was written on the backboard). The cuckoo doors have been recarved, as the originals were missing, and the minute hand is wrong but could be recarved to match the hour hand. There are individuals in the U.S.A. who carve or match bone hand patterns.

The wood plate, eight-day, double fusee movement with original bellows. The bellow tops have been re-covered.

Cuckoo and quail clock made by Johann Baptist Beha. Fifty-hour spring driven movement. Beautifully carved from walnut. Bone hands. Circa 1875.

The fifty-hour wood plate, brass gear movement made by Beha. Original birds, pendulum and bellows. This clock will give the call of the quail every 15 minutes and the cuckoo at the hour.

The finely carved doors of the Beha clock.

The vine leaf motif of the Beha clock carved in walnut.

Johann Baptist Beha of Eisenbach 115

Shelf cuckoo and quail clock with wood plate, brass gear, fifty-hour movement. Small cross at the top of the roof gable with intricately carved leaf, vine and berry motif. Bone hands and numerals. Made by Johann Baptist Beha. Circa 1880.

Shelf cuckoo clock with wood plate, brass gear, fifty-hour movement, also made by the Beha factory and similar to the cuckoo and quail clock above. The hands would have been more intricately carved from bone. Circa 1880.

Eight-day shelf cuckoo with inlay and carved eagle by Johann B. Beha. The case measures 25 inches in height. This particular example has the door for the music box at the base of the back of the clock, however the music box was optional and this clock does not have one, although it could have been fitted up very easily. Circa 1870.

116 Black Forest Clocks

Gothic style shelf cuckoo with spring driven fifty-hour movement. Clock was made by the J.B. Beha factory in Eisenbach. The crest, or gable, carving is original; on many clocks it is missing as it was usually held in place by screws. The top could easily be removed and was often lost in transport. Circa 1870.

The wood plate fifty-hour movement with stop gears missing at the bottom. The bellows have been recovered correctly, with the original pipes intact.

A spring driven wall cuckoo clock with the sleeping hunter and dog motif made by the Beha factory. The front shield is carved from one piece of wood. Fifty-hour, wood plate movement. Circa 1870. The hands would have been a fancier carved style made from bone.

Johann Baptist Beha of Eisenbach 117

Small wall cuckoo with intricate leaves, vines, and berry motif probably made by the Beha factory. The weights are carved from wood and are lead filled. The hands are not original. Wood plate, brass gear thirty-hour movement. Circa 1875.

Beha shelf cuckoo number 509 with an eight-day, double fusee, wood plate movement. The hour hand is missing and both hands should have been more intricately carved from bone. Butterfly on the roof gable.

Cuckoo and quail wall clock with thirty-hour wood plate, brass gear movement. Bird at the top and on the left. Clock is similar to the Beha model 671 and measures about 23 inches long. Bone hands and numbers, although the hands were probably more intricately carved from bone. This certainly is a Beha clock.

118 Black Forest Clocks

An eight-day, double fusee, wood plate shelf cuckoo with two birds and squirrel motif, made by J.B. Beha. The model number on the back board is illegible.

Closeup of the bird and cuckoo door at the top of the clock.

Front view of a Beha eight-day, double fusee movement. Notice levers under bird assembly. 9⅝ inches tall.

Closeup of a shut off lever in the off position. Now bird is disengaged, as are the cuckoo lift levers. When lever is returned to horizontal position, bird is engaged with its vertical lever, and lift pins for cuckoo are brought forward to engage with the lift pins. Upper level that is shown in the horizontal position is to control the gong hammer. The gong hammer arbor is engaged with the lift pins while held in this forward position by the lever. The shut off levers will allow one to sleep at night and not be awakened by the call of the cuckoo.

Side view of barrel and fusee of the same movement.

Side view of the Beha movement looking toward the front plate.

Number 509 shelf cuckoo clock made by J.B. Beha with vine leaves and a small butterfly on the roof gable. Height of the clock is about 16 inches by 10½ inches wide. Wholesale price of the clock was $11.58 at the Philadelphia Exhibition, 1876. (Dr. Wilhelm Schneider, Regendorf, Germany).

Number 671 J.B. Beha cuckoo and quail wall clock with vine leaves and berries motif, about 23 inches tall by 18 inches wide. Wholesale price was $9.00 at the Philadelphia Exhibition in 1876. (Dr. Wilhelm Schneider, Regendorf, Germany).

Number 539 shelf cuckoo clock with ferns, two eagles, deer, and running dog motif. 23½ inches tall by about 17 inches wide. Wholesale price was $17.16 at the Philadelphia Exhibition. (Dr. Wilhelm Schneider, Regendorf, Germany).

Number 312 wall cuckoo clock with larger vine leaves and berries with a bird on the roof gable; measures 16½ inches tall by 12 inches wide. Wholesale price was $5.33 at the 1876 Philadelphia Exhibition. (Dr. Wilhelm Schneider, Regendorf, Germany).

Number 512 shelf cuckoo with vine leaves and berries, carved bird on the roof gable, carved roof tiles found in both beechwood and walnut cases. Height is about 18 inches by 12 inches wide. Wholesales price was $13.86 at the 1876 Philadelphia Exhibition. (Dr. Wilhelm Schneider, Regendorf, Germany).

An unrestored Beha cuckoo clock with fifty-hour, wood plate movement. Part of the carving at the top and two finials were added to complete the restoration of this clock.

The backboard, gong, repair dates and no. 161.

Number 312 cuckoo clock made by Johann Beha, nicely carved with vine leaves and a bird on the roof. The box case was made from oak. Bone hands. The clock is about 16½ inches high and originally sold for the equivalent of $5.33 at the Philadelphia Exhibition, 1876. Wood plate, thirty-hour movement with brass gears and steel arbors.

The split cuckoo doors which are characteristic of a Beha clock.

An intricately carved pair of bone hands used for a Johann Beha shelf cuckoo.

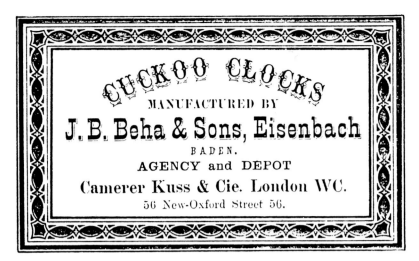

The business card of the Beha cuckoo clock factory in Eisenbach. (Antique Clocks Publishing Archive).

Award for Johann Baptist Beha at the Art and Trade Exhibition in Karlsruhe in 1877. Silver medal award. (Antique Clocks Publishing Archive).

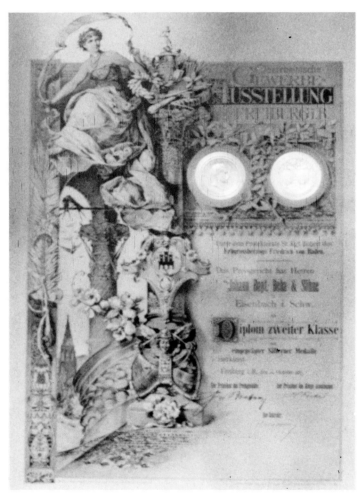

An award for Johann Baptist Beha & Söhne—Diploma Second Class—silver medal award. Eisenbach im Schwarzwald. Freiburg im Breisgau 1880. (Antique Clocks Publishing Archive).

CHAPTER 4

The Clockmakers of Furtwangen

A. MAYER, Uhrenfabrik Schönenbach bei Furtwangen

Andreas Meyer 1787-1864 first was a shoemaker, then a clockmaker. His son, German Mayer (1815-1896), after dealing clocks in England for seven years, came back to his father's workshop in Schönenbach in 1845. Three other sons of Andreas' also dealt clocks in England. In 1864, after the death of Andreas and his oldest son Joseph, German took over his father's shop. German had ten children, and three of his four sons also went to England to deal in clocks.

German's second oldest son, Joseph (1853-1903), stayed to help him make clocks. In 1882 Joseph took over and developed new types of clocks, including musical clocks with moving figures.

In 1897 Joseph became ill and could no longer work; he turned the company over to his brother Frans-Karl, who had just come back from England. Joseph Mayer died in 1903 at the age of fifty. His wife Theresia helped run the company after his death, and it was turned over to their son Alois in 1910. Alois had worked at the company previously and had played a large part in the modernization of the company in 1906, which made the workers labor a great deal easier. His sons, Heinrich and Konrad Mayer, joined the company in 1948 and 1949. In 1956 Alois withdrew from the company and public stock became available, with Alois's sons holding the majority of the stock.

The company A. Mayer O.H.G. Uhrenfabrik Schönenbach is still in business today, exporting clocks to many countries and producing floor-standing clocks after the old original B.F. clocks.[1]

LORENZ FURTWANGLER SOHNE (LFS) Furtwangen

Lorenz Furtwängler (1807-1866) was the founder of this famous and influential manufacturer of large clocks (massiv Grossuhren) in Germany. Lorenz was the son of a fruit dealer and came from a very large family, rich with children. The Furtwängler family had many notable personalities, including Professor Adolf Furtwängler and his son, a musical director, and Professor Wilhelm Furtwängler (1886-1954). Lorenz and his family lived in a small house in Gütenbach.

Lorenz Furtwängler learned about clockmaking from his older brother Johannes (born 1797). Johannes had his workshop in the family home. After Lorenz married in 1836 he established his own workshop in the family home. In 1839 he moved his family to a farmhouse called Schwefeldobel, between Neukirch and Gütenbach. Schwefeldobel was the name of a valley and politically belonged to Neukirch and religiously to Gütenbach. From this point on, he was very successful with his small company and earned a few awards at exhibitions. He became a well-known and loved personality. He was very popular and his advice was respected and welcome. For this reason he also became active in the clock trade union, later becoming a leader in the Black Forest area.

Lorenz died in 1866 and his sons, Gustav Adolf (1839-1905), Karl Hektor (1840-1911), Julius Theophil (1843-1897), and Oskar (1850-1908) continued to run the business. Two years after their fathers' death they decided on larger quarters in Furtwangen, under the name Lorenz Furtwängler Söhne (LFS). Around this time they made small weight and chain driven 8-day clocks and Schottenuhren. They made their own cases and had the same number of workers in the mechanical section and the case factory. They shipped completed clocks directly to dealers. They also made picture frame (Rahmenuhr) and box (Kastenuhr) clocks in Biedermeier styles. Eventually they also made first-class, quality large clocks called "massivuhren." In 1870 they had 28 employees.

Advertisement from the clock factory Lorenz Furtwängler & Sons in Furtwangen. Makers of the best and finest clocks. Started in 1836 (LFS). (From the book *Uhren 1913*, Prof. Dr. R. Muhe, permission German Clock Museum, Furtwangen, Germany).

Furtwangler clocks were nicknamed Schmitterlenz; Schmitterlenz was a nickname for Lorenz. These were special 8-day clocks.

Lorenz Furtwängler Söhne kept expanding and decided to go public with stockholders. On November 4, 1895 it was renamed Uhrenfabrik vormals L. Furtwangler Söhne Aktiengesellschaft Furtwangen (Bad. Schwarzwald) and was known as LFS. By 1898 they employed 143 workers. Director Georg Stehling (1868-1929) made many changes after 1900 in relation to the production of precision regulators, encouraging the production of many different styles. He was a respected specialist in the manufacture of the larger clocks.

By 1925 the company employed 500 workers. Georg Stehling was given much credit for making the company internationally known and for taking it safely through World War I. This company never produced cheaper, mass produced clocks, always staying, instead, with the production of quality clocks. The company had trouble keeping pace in the late 1920s, and with the Great Depression went bankrupt and out of business. LFS products helped to establish the Black Forest clock's worldwide reputation for quality. Its trademark was registered on September 24, 1919 with LFS in a circle.[2]

BADISCHE UHRENFABRIK, Furtwangen

This company was comprised of several small clock companies who decided to merge together. Two of the older factories to join the merger were Adam Fehrenbach and J & C Rombach. Adam Fehrenbach was already well known in Gütenbach as a maker of chain wheels and metal and iron chains for the clock industry. He started this chain factory in 1844 and moved to Furtwangen in 1858.

Founders of the J & C Rombach firm were Carl and Johann Rombach, who were cousins. They started business in 1877 in Furtwangen where they made clocks patterned after the American system. In the same year, 1877, August Rombach, brother of Carl Rombach, married the daughter of Adam Fehrenbach. This is how he got started in the clock business. About a year later Adam Fehrenbach died. Because of family relationsships, it was decided to combine the two companies in 1883 and they were named Uhrenfabrik Furtwangen. The owners were Carl Rombach, Johann Baptist Rombach, August Rombach, and Felix Ketterer (the son-in-law of Adam Fehrenbach). Felix Ketterer was also part owner of the firm B. Ketterer & Sons in Furtwangen. At this point in time, they expanded and modernized the clock factory, being the first in Furtwangen to generate power by a steam engine.

The very respected firm of Leo Faller am Bach also joined this factory in 1889. When Leo Faller's son Friedrich joined his firm in 1882 he had introduced the American system to its manufacturing process. When the Fallers joined Uhrenfabrik Furtwangen in 1889 the factory name

Front view of the piggyback movement.

An LFS wall clock. The walnut case measures 44 inches tall and is original to the finial. The clock was purchased in France. Fourteen-day piggyback movement, that is, two movements for the clock. The bottom movement is a solid plate time-and-strike type. A solid plate movement for the Westminster chime is attached to the top of the time-and-strike movement. The Westminster chime is wound at the 12.

An LFS piggyback type movement; the bottom time-and-strike for the hour, and a Westminster chime movement at the top. Rear view.

An ornate time-and-strike floorstanding clock which measures just under eight feet tall. Made by the LFS factory. Circa 1895.

The dial and top of the LFS clock.

Closeup of the weights and pendulum bob and fine craftsmanship on the case.

An LFS wall clock with brass dial and fourteen-day time-and-strike movement. Walnut case with ornate pendulum. The top is not removable on this clock. Circa 1890.

Hall clocks from the LFS factory about 1900-1920. (Gerd Bender, *Die Uhrmacher des hohen Schwarzwaldes und ihre Werke, Band II*, Verlag Müller, private catalog.)

was changed to Badische Uhrenfabrik Akteingesellschaft in Furtwangen.

The Leo Faller factory building was kept intact as a satellite factory building, and after the last merger they expanded the export of clocks and opened sales branches in Zürich and Mailand.

They also opened a factory in Hong Kong, China and in 1900 they employed 150 Chinese people, making clocks that were geared toward the Chinese market. Carl Rombach ran this factory.

The Badische Uhrenfabrik had 760 employees and manufactured 2300 clocks a day in 1902, and were considered the biggest clock company. In 1978 the factories name was BADUF and still located in Furtwangen with another branch in Simonswald, making all types of modern clocks.[3]

GORDIAN HETTICH SOHN UHRENFABRIK, Furtwangen

This company was started in the 1880s by Hermann Hettich, the oldest son of Gordian Hettich (1825-1900). Gordian, had been a salesman in a field unrelated to clocks, but his son named the company after his father because he was a very respected and famous man in Furtwangen. Gordian had a general store in his house, next to the Catholic church in Furtwangen. He added a clock packaging business where he packed clocks for shipping. He received an award in Villingen in 1858 for his efforts in getting clocks to the retail market, but he was best known for his business dealings that were unrelated to clocks.

128 Black Forest Clocks

An LFS mantel clock. Oak case that measures about 24 inches tall with brass decoration. Silvered dial. Fourteen-day time-and-strike movement. Circa 1890. (Gerd Bender, *Die Uhrmacher des hohen Schwarzwaldes und ihres Werke, Band II*, Verlag Müller, private catalog.)

An advertisement for the Badische Clock Factory in Furtwangen. (From the book *Uhren 1913*, Prof. Dr. R. Muhe, permission German Clock Museum, Furtwangen, Germany).

Freeswinger style wall clock made by the Badische Uhrenfabrik in Furtwangen. Embossed center dial piece and pendulum bob. Solid plate fourteen-day movement. Circa 1900.

Not many details are known about the clock factory except at the turn of the century. The company was called Gordian Hettich Sohn, had 64 employees, and was one of the leading clock factories in Furtwangen.

Hermann built a large factory with a carpenter shop and mass produced wall and mantel clocks. He also made, on a smaller scale, Wachteluhren, cuckoo, trumpeter clocks and automated clocks that were used in display shop windows. At the turn of the century the company changed ownership. The new owner, Max Roder, later became a director for the LFS factory; while there he kept the Gordian Hettich name. By 1920 the clock factory building was dissolved and out of business. The building was turned into a living quarters and later purchased by LFS.[4]

UNION CLOCK CO., Furtwangen

In 1870/71 Union Clock Company was a small clock case factory, run by August Weisser in Zinken Schützenbach. In 1882 Joseph Villing and Rudolf Fehrenbach took over. The name of the company was Villing & Fehrenbach. Rudolf Fehrenbach quit a year later due to poor health. Felix Trenkle took his place and the name of the company changed to Villing & Trenkle. In 1885 a London based company by the name of Merzbach, Lang & Fellheimer bought it and changed the name to Union Clock Co., and it was run by Jakob Fellheimer. It was known in Furtwangen as Fellheimer's Factory. They mainly produced inexpensive American type clocks, and by 1900 were the largest industrial company in Furtwangen.

Around 1905 Jakob Fellheimer tried to move the company, unsuccessfully. It went out of business in 1910.[5]

A three-note shelf trumpeter clock made by Gordian Hettich and Son. The movement is stamped G.H.S.

The pinned brass movement stamped G.H.S. with metal pipes and painted trumpeter figure.

130 Black Forest Clocks

Plain box clock manufactured by the Badische Clock Factory around 1910. The only identification on the clock is a B on the movement mount. Eight-day movement. Plain glass. Hour and half strike on a gong.

Gordian Hettich and Son in Furtwangen. Specialists in a variety of eight-day cuckoo clocks. (From the book Uhren 1913, Prof. Dr. R. Muhe, permission German Clock Museum, Furtwangen, Germany.)

LORENZ BOB, Furtwangen

Lorenz Bob was a very important part of the clock industry in the 19th century, being the leader in the making of mantel clocks. He was born August 10, 1805 in Dauchingen. In 1828 he started his own clock business. Up to 1840 he produced primarily eight-day clocks that were very popular for export to North America. He also made figure, nightwatchman and other types of clocks that were very respected by the clock dealers.

By 1840 Lorenz Bob started making spring driven mantel clocks that ran thirty hours and were built after the Vienna type clocks. His talents later led him to produce eight-day spring driven clocks and regulators.

In 1850 he became a teacher at the clockmaker school in Furtwangen, and was the main instructor on the subject of mantel clocks. He worked with the school until it closed in 1863. He also worked to improve the tools and machinery so they could keep up with foreign competition on the making of the clocks.

Lorenz was honored with a medal in Munich in 1854 and in Villingen in 1858 for his part in bringing the clock industry in the Black Forest to a healthy state. He died on July 6, 1878; his son Victor Bob inherited his father's talents and produced many quality clocks.[6]

Footnotes

[1] Gerd Bender, *Die Uhrmacher des hohen Schwarzwaldes und ihre Werke, Bd.II.* pp. 136-138, Verlag Müller, Villingen, 1978.

[2] Gerd Bender, *Die Uhrmacher des hohen Schwarzwaldes und ihre Werke*, pp. 180-192, Verlag Müller, Villingen, 1978.

[3] Gerd Bender, *Die Uhrmacher des hohen Schwarzwaldes und ihre Werke*, pp. 126-131, Verlag Müller, Villingen, 1978.

[4] Gerd Bender, *Die Uhrmacher des hohen Schwarzwaldes und ihre Werke*, pp. 131-136, Verlag Müller, Villingen, 1978.

[5] Gerd Bender, *Die Uhrmacher des hohen Schwarzwaldes und ihre Werke*, p. 125, Verlag Müller, Villingen, 1978.

[6] Gerd Bender, *Die Uhrmacher des hohen Schwarzwaldes und ihre Werke*, pp. 347-351, Verlag Müller, Villingen, 1978.

A Black Forest Apostle clock. The twelve apostles appear at the hour, moving through the two doors above the dial. Movement is stamped G.H.S. (Gordian Hettich Son). Missing spiral at the top and finial at the top left.

The twelve apostles and movement. The bars were added to hold the figures in line, since they tend to move about on the track.

CHAPTER 5

The Clockmakers of Triberg and St. Georgen

JAHRESUHREN-FABRIK GmbH, August Schatz & Söhne, Triberg

Four hundred-day clocks (unusual running duration) were not invented in the Black Forest. Those types of clocks were famous in the clock region of the U.S., Connecticut and New Jersey, in the middle of the 19th century. Aaron D. Grance of Newark, NJ, obtained the first patent in 1841 and Samuel Terry of Connecticut received a patent in 1852.

In 1882 August Schatz produced a series of those clocks in Triberg. They were patterned after a sample clock made by Anton Harder.

Before Schatz, Anton Harder offered his design to the clock company A. Willmann & Co. and to Gustav Becker. Both tried a few examples but found the production too complicated. August Schatz improved the sample and by 1882 he was able to deliver "Year Clocks" on a regular basis. After learning the trade from Erhard Emmler, August started his own clock company with five other clockmakers: Gerson Wintermantel, Joseph Schöpperle, Karl and German Kienzler, and Albert Fehrenbach. They bought tools and equipment from the widow of Michael Bob, who had gone bankrupt. The company was called Firma Wintermantel and Co., and later Jahresuhrenfabrik GmbH Triberg. After all of the original owners sold their shares out and only August Schatz and his sons were left (after 1923), they finally changed the firm's name to Jahresuhren Fabrik Gmbh, August Schatz and Sons—which, in addition to the "year clock," made a variety of large clocks.[1]

PHILLIP HAAS AND SOHNE, St. Georgen

Phillip Haas (1802-1874) owned a Hirschwirt (inn), and was a clock packer, or one who shipped clocks, in St. Georgen. The clock factory of Phillip Haas was actually started in 1867 by his sons Karl (1835-1900) and Ludwig Haas (1835-1904) who were twins. Phillip Haas and Sons was the first Black Forest clock factory to manufacture clocks patterned after the American system. They were one of the largest and had their own carpenter shops where they made clock cases.

Phillip Haas and Sons was famous for the manufacture of the Potato Eater. The animated wood figure at the top of the clock moved his eyes with the swing of the pendulum and lifted his arm, holding a fork to his mouth. The mouth opened to accept the potato from the plate of potatoes on his lap.

They made a variety of other clocks, from simple alarms to floorstanding clocks, and exported them worldwide to Canada, France, Russia, Brazil, South America, Africa, and other countries.

Phillip Haas and Sons won first place prizes at clock exhibitions in Vienna in 1873, Philadelphia in 1876, and Paris in 1900.

By the turn of the century they employed about 200 workers. The original founders of the company (Karl and Ludwig) owned it until July 1, 1897, when their sons Karl, Gustav, Ludwig and Albert Haas took over the business.

The factory was closed in the late 1920s, during the Depression.[2]

MATHIAS BAUERLE GmbH, St. Georgen

Mathias Bäuerle (1838-1916) started his business in 1863, beginning in an old Black Forest farm house where he made mainly shield clocks. In 1869 he built a house with a workshop in it and started to make large floorstanding clocks with large intricate movements that were referred to as "massivuhren."

In 1888 he moved the company to St. Georgen where he manufactured more delicate clocks and repeater chiming clocks; by 1889 he had 31 employees. His "massivuhren" were well respected by his fellow business contacts.

The Clockmakers of Triberg and St. Georgen 133

Shelf cuckoo number 34967 made by Phillip Haas and Son, referred to as the Hirschfall style, standing 24 inches with a 14-day spring driven movement. It originally sold in 1875 for 52 marks.

The 14-day spring driven movement by Phillip Haas.

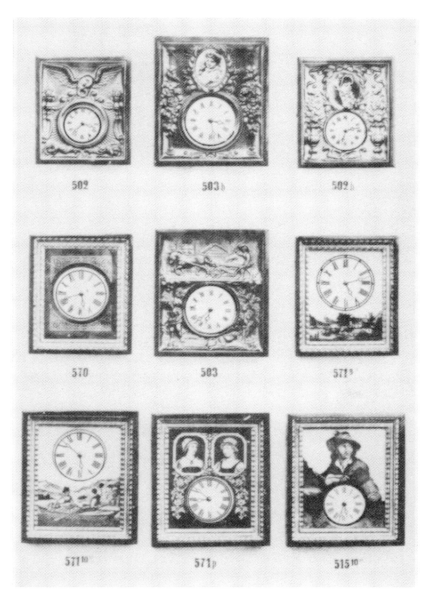

Catalog page from Phillip Haas and Sons depicting picture frame clocks, some with reverse painting on glass, some that are embossed and some with enameled dials. (Antique Clocks Publishing Archive).

134　Black Forest Clocks

Phillip Haas and Sons wall cuckoo clocks, ships style clocks, mantel, regulator, and two figure animation clocks. (Antique Clocks Publishing Archive).

Phillips Haas and Sons wall or shelf cuckoo clocks. (Antique Clocks Publishing Archive).

In 1900 his three sons, Tobias, Fridolin, and Christian took over running the company. They expanded their business to include the manufacture of adding machines in 1903 and became world famous under the name of Peerless and Badenia. They won a grand prize award in 1904 in St. Louis, Missouri. In that year they were producing 5000 first-class quality clocks per month. They made many styles of clocks which included regulators, wall clocks, floorstanding and tubular chime clocks. In 1912 the manufacture of adding machines took precedence over clockmaking, but in 1929 they were still making clocks on a smaller scale. In 1975 the company was sold to an office machine company.[3]

TOBIAS BAEUERLE, St. Georgen i. Schwarzwald

This company was started in 1864 by Tobias Baeuerle. He came from an old established clockmaking family in the Black Forest. He did business with Christian Müller and in 1864 they made a simple Schottenuhr. By 1870 they had 10 employees. In 1871 Tobias moved to St. Georgen and expanded the factory.[4]

ANDREAS MAIER, St. Georgen

This company was started in 1852 by Andreas Maier, making shields with his sons Albert, August, and Andreas. Later he also made clock cases.

In 1890 they started to make the whole clock and in 1898 had 85 employees. Since 1931 they have specialized in hands for all types of clocks.

In 1952 Andreas Maier was the largest maker of hands in all of Germany, having 100 employees and producing 12 million hands a year. Today they make precision parts for all types of items, including electronics.[5]

Footnotes

[1] Gerd Bender, *Die Uhrmacher des hohen Schwarzwaldes und ihre Werke*, pp. 138-141, Verlag Müller, Villingen, 1978.

[2] Gerd Bender, *Die Uhrmacher des hohen Schwarwaldes und ihre Werke*, pp. 142-145, Verlag Müller, Villingen, 1978.

[3] Gerd Bender, *Die Uhrmacher des hohen Schwarzwaldes und ihre Werke*, p. 146, Verlag Müller, Villingen, 1978.

[4] Gerd Bender, *Die Uhrmacher des hohen Schwarzwaldes und ihre Werke*, pp. 146-150, Verlag Müller, Villingen, 1978.

[5] Gerd Bender, *Die Uhrmacher des hohen Schwarzwaldes und ihre Werke*, p. 150, Verlag Müller, Villingen, 1978.

Catalog page from Phillip Haas and Sons depicting shield clocks, a shield animation clock, postman clocks and a postman with calendar clock. (Antique Clocks Publishing Archive).

136 Black Forest Clocks

Phillip Haas and Sons Schotten cuckoo clocks. (Antique Clocks Publishing Archive).

Heavily carved Westminster chime clock. Gothic style. The clock chime strikes on wire gongs. Porcelain dial. Glass only in the bottom part of the door. Made by the Peerless Clock Factory, St. Georgen in the Black Forest around 1900.

Two wall clocks, a number 14 Regulator and a number 15 wall calendar clock, catalog advertisement from the Tuetonia Clock Factory about 1880 in St. Georgian. The glass in the door is painted black behind a gold colored decal. (Antique Clocks Publishing Archive).

CHAPTER 6

The Villingen Clockmakers

C. WERNER, UHRENFABRIKATION

This factory was started in 1826 by Johann Nepomuk Noch of Villingen. It was first a dealership for iron and metals and also a wholesaler for clocks. Noch's son, Heinrich, helped to start the clock company. In 1861 the daughter of Heinrich Noch married Carl Werner (1832-1890). After years of travel, Werner had obtained enough experience and knowledge to take over the business in 1857 after his father-in-law died. They dealt mainly in iron and metals and, on a less active level, were wholesale dealers of clocks. However, Carl Werner was more interested in making the clocks rather than in dealing in them.

Carl eventually moved to larger quarters and shared a building with Ferdinand Meyer, who made glass frames. The production of clocks was started within these new quarters. They specialized in floorstanding clocks and expanded again in 1884 and 1895, and in 1899 they added their own case factory. Carl Werner died in 1890 and his sons, Carl and Hermann, who had already worked with him for a few years, took over the company. At this time, the company reached its highest level of success and exported worldwide. To avoid problems with customs, they opened branches in Innsbruck, Austria, France, Italy, and Warsaw, Poland. They made popular clocks based on the American method of manufacturing. They also manufactured other machines including adding machines and meters for horse-drawn coaches and taxis. They were not successful with the latter at this time, but success did come later.

In 1899 Hermann Braukmann in Villingen bought the clock factory. By 1908 they employed 850 workers, but in 1913 had financial problems because of the economy (non-business related) and they were relieved of all obligations. Schlenker and Kienzle took the company over and called it Schwenninger Uhrenfabrik Schlenker & Kienzle, and later Kienzle Uhrenfabrik A.G., Schwenningen. They ran two factories in Villingen. The known trademark in 1881 and 1887 was a horseshoe with CW in the center of it, and in 1908, the stamp of Verna.[1]

GEBRUDER WILDE
UHRENFABRIK and Unmarked Calendar Clocks

There is very little history available today about this clock factory, however it is known that the clock factory Gebrüder Wilde was founded by Leopold and Constantin Wilde in 1872. The main product of this company was the calendar clock called "System Wilde", which was very successful. The construction was developed by Leopold Wilde (born 1847) around 1876. This automatic calendar clock was used mainly in offices, and showed the time and date. The clock fulfilled the functions that a digital clock of today performs.

C. Werner in Villingen. The ad depicts calendar clock number 2500, 30 inches tall with 14-day running movement. (From the book *Uhren 1913*, Prof. Dr. R. Muhe, permission German Clock Museum, Furtwangen, Germany).

An unmarked German calendar clock, wall type with 14-day time-and-strike movement. The calendar is wound once a month. The case is walnut veneer with black trim and measures 45 inches long. RA pendulum.

An unmarked German calendar clock. The top finial is probably not original. The case measures 36 inches and is walnut veneer with black trim. 14-day movement with 30 day calendar.

CHAPTER 6

The Villingen Clockmakers

C. WERNER, UHRENFABRIKATION

This factory was started in 1826 by Johann Nepomuk Noch of Villingen. It was first a dealership for iron and metals and also a wholesaler for clocks. Noch's son, Heinrich, helped to start the clock company. In 1861 the daughter of Heinrich Noch married Carl Werner (1832-1890). After years of travel, Werner had obtained enough experience and knowledge to take over the business in 1857 after his father-in-law died. They dealt mainly in iron and metals and, on a less active level, were wholesale dealers of clocks. However, Carl Werner was more interested in making the clocks rather than in dealing in them.

Carl eventually moved to larger quarters and shared a building with Ferdinand Meyer, who made glass frames. The production of clocks was started within these new quarters. They specialized in floorstanding clocks and expanded again in 1884 and 1895, and in 1899 they added their own case factory. Carl Werner died in 1890 and his sons, Carl and Hermann, who had already worked with him for a few years, took over the company. At this time, the company reached its highest level of success and exported worldwide. To avoid problems with customs, they opened branches in Innsbruck, Austria, France, Italy, and Warsaw, Poland. They made popular clocks based on the American method of manufacturing. They also manufactured other machines including adding machines and meters for horse-drawn coaches and taxis. They were not successful with the latter at this time, but success did come later.

In 1899 Hermann Braukmann in Villingen bought the clock factory. By 1908 they employed 850 workers, but in 1913 had financial problems because of the economy (non-business related) and they were relieved of all obligations. Schlenker and Kienzle took the company over and called it Schwenninger Uhrenfabrik Schlenker & Kienzle, and later Kienzle Uhrenfabrik A.G., Schwenningen. They ran two factories in Villingen. The known trademark in 1881 and 1887 was a horseshoe with CW in the center of it, and in 1908, the stamp of Verna.[1]

GEBRUDER WILDE UHRENFABRIK and Unmarked Calendar Clocks

There is very little history available today about this clock factory, however it is known that the clock factory Gebrüder Wilde was founded by Leopold and Constantin Wilde in 1872. The main product of this company was the calendar clock called "System Wilde", which was very successful. The construction was developed by Leopold Wilde (born 1847) around 1876. This automatic calendar clock was used mainly in offices, and showed the time and date. The clock fulfilled the functions that a digital clock of today performs.

C. Werner in Villingen. The ad depicts calendar clock number 2500, 30 inches tall with 14-day running movement. (From the book *Uhren 1913*, Prof. Dr. R. Muhe, permission German Clock Museum, Furtwangen, Germany).

138 Black Forest Clocks

Advertisement for the Wilde Brothers in Villingen, Black Forest, makers of Regulators, wall clocks, calendar clocks, etc. (From the book *Uhren 1913*, Prof. Dr. R. Muhe, permission German Clock Museum, Furtwangen, Germany).

The Gebrüder Wilde gong holder.

Calendar clock made by Gebrüder Wilde. The case measures 43 inches in length and is 2 tone with walnut veneer; the finials, columns, etc., are a darker shade of brown. Enamel dial and RA pendulum insignia. The calendar movement shows the day of the week, the date of the month, and the month in the Spanish language. Germany introduced this masterfully complicated but totally efficient perpetual calendar movement. It shows the date of the month in a window, the numbers printed on a silk tape attached to a drum. The calendar mechanism is wound once a month. The date and day change their displays around midnight and the mechanism compensates automatically for months of varying length and leap year every four years in addition to keeping perfect time. These movements are living mementos of the mechanical genius of the 19th century Black Forest clockmakers.

The frontside of the calendar movement. The lever to the left of the dial is used for manual advance of the calendar.

The backside of the 14-day time-and-strike movement with the stamp of G.W. This calendar movement is wound once a month. The tape display rewinds automatically on the last day of the month and the mechanism also adjusts for the leap year.

Gebruder Wilde Uhrenfabrik was a small factory, employing about 35 workers in 1898. Alfred Wilde, the son of Leopold Wilde, took over the management of the firm prior to World War I. On May 2, 1902, at the age of 56, Constantin Wilde died after suffering a stroke. When Alfred Wilde was killed in 1915 in World War I, his father Leopold no longer felt that he could run the company in a successful manner and liquidated it.

In 1916 a man named Hiller bought some of the buildings and equipment, and he temporarily continued manufacturing on a small scale.

The movements in these clocks are a solid brass plate type with brass gears and solid steel pinions and are very reliable. The Wilde brothers should have gained a good reputation while producing these clocks. The only problem in repairing them today is that the silk tape that shows the day of the month is usually torn, worn, and shredded in some places. Sometimes a new one has to be made. The case styles of the clocks varied from the very ornate case to the simpler RA "Vienna" style. Mantel calendar clocks were also made by this factory.[2]

AUGUST NOLL

August Noll's "Kunstuhr" was a forgotten clock, lying in many pieces in an attic in Villingen. Ernst Blessing (an organ builder) started to restore it in 1950. It took three years, but with his expertise and patience, the clock was renovated. Ernst was in his 80's at the time and there were no written instructions or plans for restoration for this clock that he could follow.

It took August Noll five years to finish his first big showpiece clock in 1888. It was modeled after the famous Cathedral of Strasbourg clock. He traveled with this clock, and, to show it off, called it the largest astronomical world clock. In 1897, he completed his second improved world clock; it took seven years to build this clock. When traveling around to show these clocks it would take four days to set the clock up for demonstration and three days to take it apart and pack it to travel onto the next showing. His clocks were always admired by royalty and political people, who acknowledged his clocks as works of art and technical wonders.

These showpiece clocks were very popular with any audience. After World War I the manufacture of these clocks ceased as more modern clocks were being mass produced. Many of these astronomical clocks were discarded and disposed of, not then being perceived as being museum quality and certainly too large for one's home. The clock shown in this chapter is in the Historical Clock Museum in Furtwangen, Germany.[3]

An unmarked German calendar clock, wall type with 14-day time-and-strike movement. The calendar is wound once a month. The case is walnut veneer with black trim and measures 45 inches long. RA pendulum.

An unmarked German calendar clock. The top finial is probably not original. The case measures 36 inches and is walnut veneer with black trim. 14-day movement with 30 day calendar.

Ad from the *German Clockmaker* newspaper. Gebrüder Wilde, the calendar clock, and his clock calendar mechanism from 1895 advertisement. (Gerd Bender, *Die Uhrmacher des hohen Schwarzwaldes und ihre Werke, Band II*, Verlag Müller, private catalog.)

UHRENFABRIK VILLINGEN AKTIENGESSELSCHAFT and UHRENFABRIK VILLINGEN J. KAISER GmbH

The once famous Villingen Uhrenfabrik was started in 1852 by the Maier brothers in a simple clockmaking workshop. In 1863 they moved to bigger quarters to set up a factory. They mainly made mantel clocks.

In 1899 the Maier brothers joined with Maurer Pfaff to do business as the Uhrenfabrik Villingen A.G. In 1900 the company of Wilhelm Jerger of Niedereschach merged with them also.

After bankruptcy in 1914, Josef Kaiser took over and rebuilt the factory and the name was Uhrenfabrik Villingen J. Kaiser GmbH. Josef Kaiser came from an old Lenzkirch family and learned the trade in the famous Lenzkirch Clock Factory. He also studied in several of the other large Black Forest clock factories. After his death in 1940, his three sons, Franz, Oskar, and Rudolf, took over this very famous company.

With hard work, these three managed to get through the depression, and after 1950, made good progress. In 1973, however, they could not keep up with their larger competitors and went bankrupt.[4]

Footnotes

[1] Gerd Bender, *Die Uhrmacher des hohen Schwarzwaldes und ihre Werke*, pp. 151-154, Verlag Müller, Villingen, 1978.

[2] Gerd Bender, *Die Uhrmacher des hohen Schwarwaldes und ihre Werke*, p. 156, Verlag Müller, Villingen, 1978.

[3] Gerd Bender, *Die Uhrmacher des hohen Schwarzwaldes und ihre Werke*, p. 642-649, Verlag Müller, Villingen, 1978.

[4] Gerd Bender, *Die Uhrmacher des hohen Schwarzwaldes und ihre Werke*, pp. 154-156, Verlag Müller, Villingen, 1978.

Clock with world times, calendar, music box, automated figures (rooster, cuckoo, night watchman, bells, and apostles). Approximately 50 moving objects overall. Made by August Noll, Villingen about 1880-1885. About 10 feet in height. (Deutsches Uhrenmuseum, Furtwangen/Schwarzwald, photo, Callwey Verlag, München, Germany).

An unmarked calendar clock, the day and month written in German. The oak case measures 43 inches long.

An unmarked shelf calendar clock with 14-day time-and-strike movement and 31-day calendar wound on the side of the walnut veneer case with brass decoration. Silvered metal dial. The calendar writing and tape have been redone with a misspelled day of the week.

CHAPTER 7

The Schwenningen and Schramberg Clockmakers

FRIEDRICH MAUTHE, Schwenningen

Friedrich Mauthe (1822-1884) was an infant when his father Jakob died. Friedrich went to grade school in Schwenningen and then high school in Bolingen. His legal guardian preferred him to learn the retail trade business and sent him to work in a bookstore in Rottweil. After his apprenticeship he travelled the countryside taking jobs to gain experience.

Friedrich did not stay away long and returned to marry the daughter of his neighbor, who was a farmer named Kienzle. After the marriage he moved to the top story of his in-law's house and they opened a small shop in the basement. It was a general store, selling groceries and fabrics for clothing. Being a man of many ideas, he decided to carry tools and parts for the local Schwenningen clockmakers. He carried brass hands, wheels, bells, and different types of iron or metal wire. He imported these parts to the Black Forest. Later, in the 1850s, he decided to become a wholesaler (uhrenpacker) for complete clocks which he shipped all over the world.

He became very involved in this venture and decided to start his own clock factory, with Jakob Haller as his partner. Jakob Haller was a respected clockmaker in Schwenningen and, being the expert, he ran the factory. Friedrich ran the sales and kept the records. At the same time he wanted to keep his wholesale business for parts and tools. He needed financial help and tried to get it from Jakob Haller's brother Johannes, but was only approved from the state for one-half of what they needed. This was disappointing to them, and they decided against any expansion.

By the end of the 1860s he had 12 workers in his shop and farmed out other work to Black Forest families. He made thirty-hour spring driven clocks, cuckoo clocks, and twelve- and twenty-four-hour Shottenuhr. In 1872 his oldest son Christian started working for him. In 1876 Friedrich retired

Round top box clock with one piece of beveled glass in the door. Black walnut stain on an oak case. Hour and half hour strike on a 3 rod chime (bimbam) which has *Divina Gong* stamped on the base. With some research it was found this clock was made by Friedrich Mauthe of Schwenningen. Circa 1912.

Wall clock made by Friedrich Mauthe, which measures 36 x 18 inches. Movement is stamped FMS.

Spring driven regulator by Friedrich Mauthe of Schwenningen. Second Baroque style case. Rod chime half hour and hour strike. 14-day movement. This clock was purchased in France. Circa 1900.

and turned the shop over to his sons, Christian and Jakob. They took over the Gasthaus zur Krone (an inn) and turned it into an industrialized factory. In 1881 they modernized with steam power, being the first in Schwenningen to do this.

Friedrich's father Jakob had inherited the famous Gasthaus zur Krone from his father Johannes. Johannes was active in the nitric acid business; this acid was produced out of cow manure and found in the local farmer's cow barns. At the time, it was a big business and the nitric acid was used to make gun powder.

Friedrich's two sons brought recognition to the factory as years passed. In later years they made all styles of clocks to include hall, wall, and mantel types. In March 1976 Mauthe closed their doors.[1]

JACOB KIENZLE, Schwenningen

Jacob Kienzle was born April 12, 1859 in Schwenningen. Jacob's father died three months before he was born. His father's sister had married Friedrich Mauthe. His legal guardians were Friedrich Mauthe and his real mother. Friedrich Mauthe also paid for Jacob's schooling and took care of him.

In 1873, when he was fourteen, Jacob already worked for a short time in his uncle Friedrich's clock business. They did not really want him to be a clockmaker, so he was placed into an apprenticeship to learn retail business in Triberg. Four years later, after he finished his apprenticeship, he worked in a fabric mill in Mühlhofen. In 1879 he came home to his uncle and worked for Friedrich Mauthe as a salesman in the shipping department. Jacob spent many Sundays writing all the bills. He was an eager learner, wanting to learn all he could. In his spare time he taught himself the French language. After working with his uncle for four years, he moved on to find more challenging work.

In 1883 Jacob went to work for Christian Schlenker, and at age 24 married Christian's daughter. His wife's grandfather Johannes Schlenker started the business in 1822, and because of Jacob's marriage, Christian made him a part owner of an established business. A brother-in-law, Carl Johannes Schlenker, was also made a partner in 1883, and the name of the business changed to Firma Schlenker und Kienzle.

Due to Jacob's enthusiasm and hard work, he helped expand the export business. His father-in-law had much trust in him and turned the farmhouse into a factory for the clock business. Only two years after Kienzle joined the company they bought more buildings, near the train station. They expanded further, and production started in these buildings in the same year that Christian Schlenker died (1885).

A German calendar shelf clock with original bracket, attributed to Kienzle/Schlenker factory, about 1890. The calendar movement shows the day of the week, date of the month, and month in the German language. The case is stained in walnut with a dark trim on the columns, finials, buttons, etc. The clock movement will run for eight days while the calendar movement is wound once a month. The dial is metal, silvered with etched numerals.

Black Forest Clocks

RA style clock with calendar mechanism. The indicated day of the month is shown in the center of the dial. Walnut case is 34 inches long. Hour and half-hour strike. 14-day movement by Kienzle. (Lyon's Antique Clocks, Costa Mesa, CA)

Small spring-driven German Regulator by the Kienzle clock factory. Metal dial with Arabic numbers. Walnut case. Gong chime. 8-day movement. Circa 1900.

View of the dial and calendar.

Big changes occurred in 1888 in the Schwenningen clock factory. Austria and Hungary hiked the custom fees for imported clocks. Because of this, the factory could not afford to export to Austria anymore, and therefore could not compete with Austrian clockmakers in Vienna. Jacob Kienzle did not want to lose this market, so they started a branch in Böhmen (Bohemia). In 1888 Carl Schlenker went to Komotou, Bohemia and launched this company. The two companies employed 120 people and in ten years that number tripled. Unlike Jacob, his brother-in-law C.J. Schlenker did not want to expand further. Jacob left the company and started his own. Jacob Kienzle was the sole owner and the new company surpassed the old. In 1905 he employed 400 workers, and was the most prominent clockmaker in the Austrian/Hungarian monarchy. He also expanded to Milan and Paris.

Jacob's brother-in-law's company had a very small and short existence. Jacob bought him out in 1899 and Schlenkers Co. became part of Kienzle. In 1919 Jacob Kienzle retired. His oldest sons Herbert and Christian took over this well respected company. Jacob never forgot where he came from; he always contributed a great deal to charities and the community. Because of that, in 1927 the city of Schwenningen honored him as a valued and honorable citizen. He was also honored at the technical school in Stuttgart with an honorary doctors degree in 1927.

Jacob died in 1935, and by that time 3,000 workers were employed at the Kienzle Uhrenfabrik. The company specialized in the production of 14-day solid plate movements. Some facts on production:
  1888 produced 65,000 clocks annually
  1892 produced 162,000 clocks annually
  1913 produced 4,500,000 clocks annually, and employed 2,000 workers
  1939 produced 5,000,000 clocks annually, and employed 3,500 workers

In their peak years they produced hall, wall, and mantel clocks.[2]

GEBRUDER JUNGHANS, Schramberg

The Junghans family lived in Horb on the Nekar River, where the name dates back to 1696. Nicholas Junghans left this town in 1815 and eventually settled in Zell, a small village where he went to work in a porcelain factory as a handyman, gaining much experience and training as a copper etcher, designer, and porcelain maker. In 1820 he married Barbara Schonenberger and eventually they had four children, Franz Xavier, Erhard, Arthur, and a daughter, Frieda.

In 1841, Erhard Junghans went to work for a

An advertisement for Black Forest cuckoo clocks from E.R. Schlenker, from the February, 1891 issue of *The Jewelers Circular and Horological Review*. (Horological Data Bank, National Association of Watch & Clock Collectors Museum, Inc.)

Wall clock made by Schlenker-Kienzle about 1890. Case measures 36 inches long. The finials at the top and the bottom should have been the fluted type also.

Junghans RA. Walnut case with black trim. Two piece porcelain dial. Hour and half-hour strike on a wire gong. 14-day movement. Circa 1900. (Antique Clock Gallery, Long Beach, CA).

Ornate spring driven wall clock made by the Junghans clock factory. Simple time-and-strike 14-day movement. Walnut case. Circa 1900.

Spring driven clock made by the Junghans clock factory. The 8 point star insignia on the dial indicates this clock was made about 1890. Twenty pieces of brass applied to the oak case. 8-day movement.

The Junghans factories from 1864 to 1898.

The Junghans factory in the year 1880. (Photos courtesy of Antique Clocks Publishing Archive).

straw manufacturing firm. They made straw hats and door mats. Erhard eventually was appointed as a management assistant and worked in Switzerland and France, gaining experience and promotions in straw weaving items, clocks, and porcelain. Later, a non-profit organization was formed, referred to as "Poorman's Corporation," which made straw hats and mats. Erhard (30 years old now) was the manager and it was suggested by the district commissioner, Ferdinand von Steinbeiss, that he should also look into starting a clock hardware business. At this time the clock industry in the Schramberg/Schwenningen area was small. The approximate 26 inhouse workshops had 130 workers making the traditional Black Forest clock (Shield type, etc.).

In 1859 Erhard Junghans and his brother-in-law Jacob Zeller purchased land in Schramberg to build an oil seed mill. Building number 43 (now number 28) occupied the land and was still in existence in 1976. It did not make a lot of money, however, and in 1861 it was converted and Erhard started to make clock parts to include hands, hinges, glass doors, wire parts, and hardware. Erhard continued to correspond with his brother Xavier who had emigrated to the United States. Erhard asked his brother to buy better tool and die machinery in the U.S. His brother did so, and it was sent to Schramberg and installed immediately. Xavier returned to Schramberg, a specialist in cabinetry. Zeller decided to resign as a partner and did so after six months, while Erhard and Xavier planned to make clocks after the American method, about 1865. The Junghans parts manufacturing plant cut moldings and veneer to specifications for cabinetmakers and made all types of clock

Junghans label on the back of the case says "Junghans clocks are best."

Junghans box clock with leaded/beveled glass in the door. Eight-day movement with half hour and hour strike on a gong. Light blonde colored case. Circa 1910.

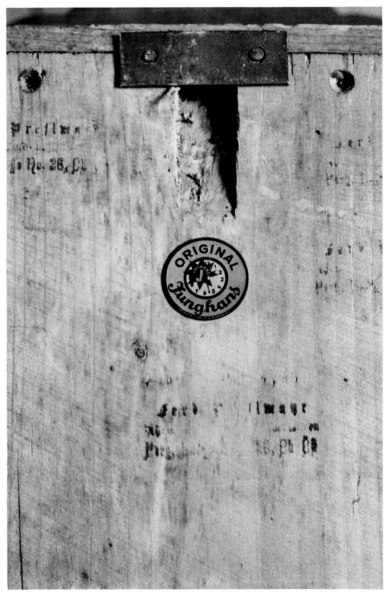

Emblem on the back of the case. Original Junghans.

The Kangaroo swinger by Junghans. Circa 1900.

The Elephant swinger by Junghans. Circa 1900.

hardware, carrying a large assortment for small or large clock manufacturers. This allowed the local farmhouse clockmaker the option to standardize their clocks, economize on parts, and speed up their production. At the same time, the straw manufacturing company flourished. In 1864/65, a report showed that Junghans had made 25,000 clock boxes which were mass produced.

Junghans initiated laws along the way to industrialization. Laziness on the job was grounds for dismissal. No alcohol or cigarettes were allowed in the factory. Workers were paid twice a month. These rules were quite different from the in-house clockmakers rules, if there were any.

Xavier's fifteen years worth of American experience helped produce Black Forest clocks patterned after the American method. At an exhibition in Stuttgart in 1866, Erhard Junghans exhibited four different styles of clocks made after the American method: a marine clock, an alarm clock, a weight driven clock, and a spring driven clock, made in wooden boxes with wood plate movements. At the same time, American-made clocks from Eli Terry and Seth Thomas were being bought by the European market and there was worry by Junghans that the entire European market would be flooded with the American clock. An American clock factory could make 450,000 clocks in 1860. In 1868 Junghans produced about 360 clocks a week, made by 72 workers.

Having grown beyond the definition of a village, in 1867 Schramberg officially received the title of City. Junghan's sons Erhard, Jr. and Arthur were preparing themselves for work in the clock industry, and Paul Landenberger was hired in 1868 to work with the company. Erhard Junghans Sr. died in 1870, only 47 years old. Paul Landenberger was to handle all business for the factory.

Erhard, Jr. and Arthur began to assert themselves in the management of the company, and Xavier returned to the United States after being informed that the factory could only be passed on to the sons of Erhard, Sr. Arthur went to the U.S. to learn the American method of clockmaking, but had no luck at first, finding it very difficult to be hired. He finally found work as a bus boy and janitor at the Winstead Clock Factory, earning $1.00 a day. He eventually worked on an American machine and made many notes on American machinery, thinking that one day he would apply it to the Black Forest industry. He returned to an expanded factory, now with four buildings. Paul Landenberger had left and opened a clock factory called Landenberger and Lang, later the Hamburg American Clock Factory, which became a strong competitor.

Erhard Junghans, Jr. was asked by his mother to take over the company; he accepted and the factory was officially transferred to the sons Erhard and

152   Black Forest Clocks

Castle style mantel clock made by the Junghans clock factory. Enamel dial. Case is 16 inches tall with brass decor. 14-day time-and-strike movement. Circa 1900.

The label on the back of the case of the Junghans clock. The clock dealers Camerer, Kuss & Co. in London.

The Cleopatra swinger by Junghans. Circa 1900.

The Schwenningen and Schramberg Clockmakers

The Barmaid swinger by Junghans. Circa 1900.

The Onion Boy swinger by Junghans. Circa 1900.

The Bat Boy swinger by Junghans. Circa 1900.

Arthur in 1876. At the time, 225 people were employed by Junghans. In 1875, the Number 10 Baby Alarm clock came on the market, was very successful, and was in production for about fifty years. In 1877 the first trademark was registered. It was an eagle surrounded by stars with the inscription E. Pluribus Unum, which means "Out of many one."

Junghans registered its first patent on April 22, 1879 and through 1920 registered about 300 patents under the Junghans name. In 1883 Junghans started their development of pocket watches, and they found a good market for them.

In 1888 the five point star insignia was used and later, in 1890, the eight point star insignia was used. By 1893 the factory reached the 1,000,000 production mark.

Junghans pursued other areas of manufacturing and designed the automatic mail stamp cancelling machines, loud speakers, alarm systems, cutting machines, etc., and experimented with many other types of machinery through the years.

In 1902 Junghans opened a branch factory in Rottenburg, Black Forest, and in 1900 a merger with Thomas Haller took place. In 1904 Junghans had production plants in Schramberg, Rottenburg, Ebensee, Austria (formerly Gebruder Resch), Venice, Paris, Schwenningen, Deisslinger, and Lauterbach.

154   Black Forest Clocks

Miniature regulators from Junghans. (Photo courtesy of Antique Clocks Publishing Archive).

A small wooden alarm clock with glass in the sides of the case, made by Junghans. Circa 1900.

Another wooden alarm clock made by Junghans. Case measures 8 inches tall. Circa 1880.

The Schwenningen and Schramberg Clockmakers 155

Mantel clock made by the Junghans factory. Time-and-strike 8-day movement. The case is 23 inches tall.

Westminster chime wall clock made by the Junghans factory. Case is made of walnut and is 38 inches in length. Dial is metal and painted silver. 8-day movement. Circa 1900.

The Junghans label on the back of the case.

Junghans advertisement. The biggest clock factory in the world. United Clock Factories from Brothers Junghans and Thomas Haller, Schramberg. (Antique Clocks Publishing Archive).

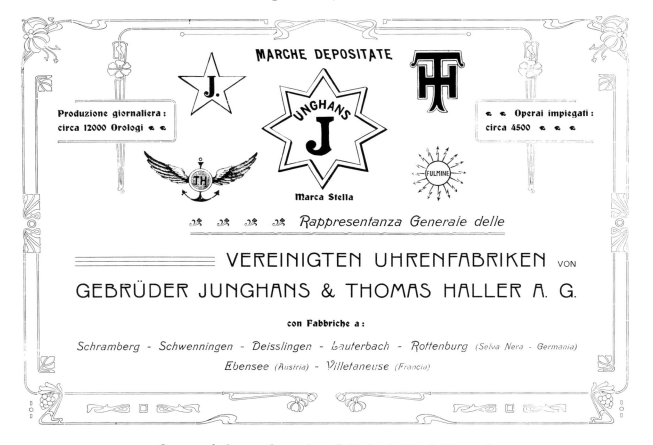

Some of the trademarks of United Clock Factories, Junghans and Thomas Haller in various cities in Europe. (Antique Clocks Publishing Archive).

The outbreak of World War I had an impact on the Junghans factory. Fifty percent of the workers were in the military and production was converted towards the war effort. Later the Junghans factory merged with the Hamburg American Clock Company and the Gustav Becker Factory in Schlesien; the Lenzkirch Factory was merged with Junghans in 1928.

Some information on Junghans factory production is provided. In 1875, 37,000 timepieces were manufactured. In 1885 the number was 254,000. In 1890, 851,000 timepieces were produced. In 1905, 2,881,000, and in 1914, 3,600,000. In 1974 approximately 5,400,000 timepieces were produced. Through these years Junghans produced alarm, mantel, wall, floorstanding, animated, calendar, musical, and cuckoo clocks.[3]

It should be said that after the introduction of American manufacturing methods by the Bros. Junghans in Schramberg, the majority of clock production shifted to the Black Forest at Württemberg. It was here that a modern and competitive large scale industry soon came to prosper. Unhindered by the old tradition it employed up to 3,000 workers by the turn of the century.

The main locations of the Württembergian clock industry were the market town of Schwenningen (pop. 12,000) in the district Rottweil and the town of Schramberg (pop. 10,000), which is just a few hours away, located in the Oberndorf district.

It was precisely Schwenningen/Württemberg, where, at the end of the 18th century, the actual "Black Forest clock industry" first moved from the neighboring part of the Black Forest at Baden. Here, it quickly caught on and gradually spread to the surrounding districts of Rottweil, Oberndorf, Speichingen, and Tuttlingen. However, Schwenningen and Schramberg always remained the main locations and real centers of the Württembergian clock industry.

The well-known entrepreneurs Johannes Bürk, Friedrich Mauthe, and Jakob Kienzle of Schwenningen, as well as the Bros. Junghans of Schramberg, are considered the founders of industrial clock manufacturing in Württemberg.[4]

HAMBURG AMERICAN CLOCK COMPANY,
Schramberg

In 1869 Paul Landenberger settled in Schramberg; he was twenty years old. Paul first worked for the Junghans Brothers Clock Factory in 1868 as a manager. Erhard Junghans died in 1870 and Paul was left in full charge of the Junghans factory; Erhard's sons were not available to run the factory. Erhard's wife seemed to hinder the further advance-

Small, spring driven wall clock made by the Hamburg American Clock Factory. The center of the dial is embossed. Eight-day movement. Circa 1900.

ment of Paul Landenberger, even though he had married her daughter Frieda. In 1875, after much turmoil Paul decided to put his own practical experience with industrial factory methods to work, and started his own clock business in Gottelbachtal. The Landenberger and Lang Clock Factory was founded with Phillip Lang.

In 1883, for economic and financial reasons, the company was changed to an open stockholder corporation, renamed Hamburg American Clock Factory, and relocated in the city of Schramberg. Paul Landenberger was the director and general manager. The company had a large export market to England and its territories, and to Scandanavia. By this time the name was known worldwide. By use of a production line, the clocks were successfully marketed throughout the world. They used their crossed arrow trademark from 1891, and beginning in 1905 used a small oil lamp, which could be found stamped on the movement or on the dial of the clock.

The factory consisted of a number of buildings, with about 70 machines to make parts and components, and employing about 150 people. They produced a large selection of clocks and the increase in sales meant an expansion of the factories, so the company purchased more land and machinery.

Adolf Holbfass, who had worked with Paul Landenberger from the beginning of the factory, resigned in 1887 and Christian Landenberger filled his position as the sales representative for the company in London. He later came back to Schramberg as manager of the company.

Again, in the years 1887-1890 sales were brisk and exceeded previous years by about a third. It was decided, again, to enlarge the plant and purchase new machinery. As time passed, HAU employed 300 workers and received awards at exhibitions in Australia and Germany. The company rewarded their workers with productivity related bonuses, which helped HAU produce a sales record in 1890-1891. With the long awaited railroad connection between Schramberg and the European railroad system, expansion came again, as Germany prospered. The company now employed over 700 workers.

A modern sawmill was set up and a new facility in Alpiersbach was opened and purchased in 1900/01. The company had branches in Austria and France, and by 1905 they employed about 1000 people. The company continuously expanded their sawmill, wood storage, and carving facilities around 1912. The case making factory was enlarged to accommodate the large selection of mantel, wall, and floorstanding clocks being produced. The work force grew to 1300 by about 1914. At the

Small time-and-strike wall clock that measures 24 inches tall, made by the Hamburg American Clock Factory. Model number 2535. Eight-day movement with paper dial. Circa 1910.

outbreak of the World War I, the company lost one half of their work force to the armed forces. It was difficult to get adequate quantities of brass, copper, and steel, because they were being used for the German war effort. The war did not stop the company, though. They worked day and night shifts and orders were filled as best as possible under the circumstances.

After the war, the company had to adjust to the peacetime production. Basic clock materials were still rationed by import quotas. Somehow, the company managed to avoid any serious setbacks and they re-opened previous export routes and worked diligently to streamline operations.

In 1925 the company ranked high in the worldwide clock factory business by utilizing the American method of clockmaking. Gears were stamped from brass and new methods of mass production of parts and movements were implemented.

The Management of HAU Co. was as follows:

Paul Landenberger, founder, member Board of Directors.

Wilhelm Deurer, Chairman of Board of Directors.

Paul Landenberger, Jr., General Manager, then controller in 1900. 1915—Board of Directors; 1921—General Manager of HAU.

Paul Gunsser, Administrative Manager in 1897 and Director in 1921.

Richard Landenberger, Director.

Kurt Landenberger, Researcher on U.S. methods of clock manufacturing. 1921—Director.

Franz Goede, Controller.

The facilities occupied an area of over 4 million square feet, with more than 2200 people on the payroll. Hundreds of different machines were used to make the clock parts, with new tooling designs constantly being developed by the company's engineers. The cabinet shop manufactured the clock cases from the best lumber, cut in their own sawmill. Daily production was placed at about 15,000 timepieces, which included hall, alarm, mantel, and wall clocks, and also special clocks for other industries. The company always worked with its dealers to supply the right styles of clocks for the current world market. The company sold clocks throughout Europe, Africa, America, Asia, and Australia.

The HAU sales organization in Hamburg, Germany made outstanding contributions to the company and helped in the worldwide distribution of their clocks.

It should be noted that in the early years of clock production, daily production was perhaps 12 clocks, but the methods in the 1880s produced thousands of timepieces daily. The HAU company employed the American method of manufacture, and this was the primary advancement over the old methods of clockmaking. The HAU clock company created more than 800 different styles for its worldwide markets.

In 1932, because of the competition of other factories, the HAU company merged with Gebruder Junghans and the HAU name disappeared, although the name Hamburg American Clock Co. was used in published catalogs until 1937.[5]

Footnotes

[1] Gerd Bender, *Die Uhrmacher des hohen Schwarzwaldes und ihre Werke*, pp. 245-249, Verlag Müller, Villingen, 1978.

[2] Gerd Bender, *Die Uhrmacher des hohen Schwarzwaldes und ihre Werke*, pp. 249-252, Verlag Müller, Villingen, 1978.

[3] Karl Kochmann, *Junghans Story*, pp. 5-88, Antique Clocks Publishing, Concord, 1976.

[4] Gerd Bender, *Die Uhrmacher des hohen Schwarzwaldes und ihre Werke*, pp. 231-234, Verlag Müller, Villingen, 1978.

[5] Karl Kochmann, *Hamburg American Clock Company*, pp. 3-90, Antique Clocks Publishing, Concord, 1980.

A small alarm clock in a wooden case with bell on top, made by the Hamburg American Clock Factory. Circa 1900.

CHAPTER 8

The Clockmakers of Neustadt, Lenzkirch, and Schonwald

UHRENFABRIK FURDERER JAEGLER & CIE AND UHRENFABRIK NEUSTADT i. SCHW. AKTIENGESELLSCHAFT [Corp.] Neustadt

The Uhrenfabrik (Clock Factory) Neustadt i. Schw., (Black Forest) Aktiengesellschaft was originally founded by the Lenzkirch trading company Fürderer Jaegler & Cie.

This Lenzkirch trading company carried clocks as its main product and had its own clock store in Strassburg, France. In the middle of the 19th century the trading contracts with France called for a particularly high import tax. The Lenzkirch traders established their own manufacturing firm in Buchsweiler/Alsace, in 1855. In 1865, new German-French trading contracts led to essential tax reductions so that more efficient manufacturing at home was given preference. Therefore, on June 30, 1865, the Lenzkirch traders set up the clock factory Fürderer Jaegler & Cie. in Neustadt and opened an outlet in Paris under the same name.

The firm Fürderer und Jäggler u. Comp. in Neustadt employed approximately 350 workers, of which about 50 were women, who either worked in polishing or in the production of clock parts. The group of male workers included 50 joiners, turners, and woodcarvers, a small number working as casters and packers, with the remaining working as clockmakers.

The main depots of the factory were located in Strassburg, Paris, Mühlhausen, Basel, Hagenau, Metz, and numbered approximately 20.

The company made only a small profit, so the large investment of capital did not pay off. On July 1, 1882, the enterprise was transformed into a corporation, but despite the corporate structure, it failed to be successful. In 1886, the trading company Fürderer Jaegler & Cie. decided to get rid of their unprofitable business. The entire property of the firm, consisting of nine buildings and the mechanical equipment, was sold for the low price of 75,000 marks.

In December of the same year, a successor company was established under the name Uhrenfabrik Neustadt i. Schw. Aktiengesellschaft, which took up production on January 1, 1887. Yet this enterprise, too, was destined to be only short-lived. At the end of December 1889, liquidation of the firm was decided, being completed by January 25, 1892.[1]

WINTERHALDER & HOFMEIER, Neustadt, Schwärzenbach and Friedenweiler

Matthias Winterhalder (1686-1743) began making clocks in the Black Forest about 1710, in the Kalte Herberg Inn in Urach. The inn was home to the Winterhalder family and was located on the north slope of a mountain, which was unsuitable for farming. Matthias Winterhalder sought to supplement his income by making clocks part-time. The inn was a gathering place for travelers and, as in any local inn, you could hear the latest news around town.

Matthias first made clocks with movements that had wooden plates and gears, with steel wire pinions, usually weight driven by a rock or a primitive-looking weight of some kind. These clocks were not the most accurate timekeepers. The clocks were sold in the inn or by the traveling salesman who carried many clocks on his back and walked or rode by wagon from each village to sell his clocks. After Matthias Winterhalder's death in 1743, his daughter Magdolena married George Winterhalder (1730-1783). Their oldest son, Thomas Winterhalder (1761—1838) sold the inn in 1811 and went to work as a full time clockmaker. He moved to Friedenweiler with his family in 1816 to the Alte Haus and was the town treasurer for ten years, from 1819 to 1829.

Matthäus Winterhalder (1799-1863), the oldest son of Thomas, bought the clock shop from his father in 1830 for 1800 gulden, which was a lot of money at the time. He did very well in the next few

Close-up of the cuckoo door and bone hands.

Nicely carved shelf cuckoo with brass plate movement. Bird and deer motif. The case is stained almost black. Bone hands. Circa 1870. The carving on the case is asymmetric and appears to indicate an earlier (or better), more detailed, style of wood carving. Many clocks carved with this intensity had wood plate, brass gear movements. This movement is stamped Furderer Jaegler.

years with his clock business. In 1850 Matthäus and Johannes Hofmeier joined as partners in the clockmaking business and formed Winterhalder & Hofmeier, Schwärzenbach, using the trademark "W & H" over "SCH". The new company had factories in Friedenweiler and Schwärzenbach.

At the time of Matthäus Winterhalder's death in 1863, his estate was estimated at 5300 gulden in cash, and 303 gulden in machinery—a good amount of money.

Johannes Hofmeier (1802-1876) was also part of an innkeeper family and used his mechanical abilities to become a clockmaker's apprentice. About once a month he or his brother took clocks to London, England to try to sell them, taking perhaps twelve clocks or so at a time. They carried these clocks on foot to Germany and Holland, then

The cast brass movement with original bellows. Stamped Furderer Jaegler Co.

162 Black Forest Clocks

The stamp of Fürderer Jaegler which could also be found stamped on the backboard of some of their cases.

Matthäus Winterhalder (1799-1863). He formed the Winterhalder & Hofmeier Clock Factory in 1850 with Johannes Hofmeier. They had factories in Neustadt, Schwärzenbach, and Friedenweiler. (Antique Clocks Publishing Archive).

Johannes Hofmeier (1802-1876), partner to Matthias Winterhalder. (Antique Clocks Publishing Archive.)

A small Winterhalder & Hofmeier shelf clock patterned after an American made cottage clock. The American system.

The Winterhalder & Hofmeier movement made of brass with bell strike.

London. However, this took time away from making clocks and eventually they used the Uhrenträgers (clock peddlers), as others did, to peddle clocks to other dealers or private buyers.

In 1850 Anton Winterhalder (1838-1912), the son of Matthäus, began an apprenticeship under Johannes Hofmeier. He eventually completed his clock studies and married his teacher's daughter, Elizabeth, in 1864. He also became a director of the company in this year.

In 1869 the company was incorporated at Friedenweiler and Schwärzenbach. Many of the clocks made by the company around this time were sent to England. When Johannes Hofmeier died in 1876, the name Winterhalder and Hofmeier was kept, reflecting both the respect Matthäus had for his partner and the solid reputation the company had built for making good clocks.

Anton Winterhalder enjoyed a productive relationship with all of his fellow workers, and successfully motivated them into making quality clocks. He worked with them, knew them by first name, paid them well, and created a high morale. The factory was enlarged and moved to Neustadt in 1896.

Many members of the Winterhalder family were involved in some form of clockmaking. Anton's older brother Thomas (1834-1906) specialized in making chime strike attachment units. Another brother, Karl (1836-1918), also worked with the company. Anton's six sons were all involved in clockmaking. Ludwig (1877-1959) and Linus

Catalog title page from the clock factory Winterhalder & Hofmeier in Neustadt, Schwärzenbach and Friedenweiler.

164 Black Forest Clocks

Winterhalder & Hofmeier clock that measures 13 x 9½ inches. Case is mahogany with glass on the sides and top of the case. Westminster chime on four gongs.

Winterhalder & Hofmeier mantel clock with inlay and quarter hour strike. The case measures 13 inches. Circa 1900.

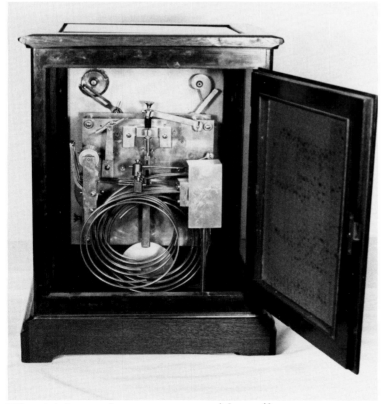

The high grade movement with strike on gongs (Westminster chime).

Winterhalder & Hofmeier mantel clock which measures 18 inches tall. Quarter strike movement. Silvered metal dial with subsidiary dials to adjust time and silent/chime.

Winterhalder & Hofmeier mantel clock which measures 14 inches and is made of oak. Quarter strike movement. Chime/silent and time adjustment at the top of the dial. Glass sides on the case.

Winterhalder & Hofmeier 8-day, half-hour and hour strike mantel clock in an oak case, which strikes on 2 round gongs. Silvered chapter ring. Circa 1900. (Antique Clocks Publishing Archive).

A Winterhalder & Hofmeier mantel clock. The case is made from oak and measures 16 inches tall. Metal, silvered dial. 14-day movement. Circa 1900.

(1878-1932) took over the company when Anton retired to Freiburg, although Anton was still involved with the company. Ludwig worked with factory production and Linus handled sales and finance. The company flourished, sending clocks to England, the U.S., Russia, Japan, and China. Another son, Anton (1872-1945), specialized in striking systems for Westminster, Whittington, and others, in Schwärzenbach. Three other sons, Johannes (1866-1935), Hermann (1876-1945), and Bernhard (1880-1958), also had small shops related to making striking clocks. Other family members were involved in Black Forest clockmaking through these industrialized years of 1870-1930. Through the years, the Winterhalder & Hofmeier factory produced quality mantel clocks for the English market, as well as wall and floorstanding clocks. The movements and cases were of the highest quality. World War I stopped production for a time and the Depression (around 1930) finally led to the demise of this factory.

W & H bracket clock with 8-day movement and quarter hour strike on 2 spiral gongs. Time adjustment on the left and chime/silent adjustment on the right above the dial. Silvered chapter ring. Circa 1900. (Antique Clocks Publishing Archive).

Other trademarks that were used include: (on two lines) "W & H"/ "SCH"; (on three lines) "W & H" / "in" / "SCH"; and (on two lines) "W & H in SCH" / "Germany". These could be found on variations of their factory clocks, which today, because of their durability, are highly collectible and very reliable clocks. The Winterhalder & Hofmeier clocks also helped to establish the Black Forest's worldwide reputation for quality clocks.[2] Their movements were the highest quality and perhaps were the best and most reliable of all that were produced in the Black Forest.

AKTIENGESELLSCHAFT FUR UHRENFABRIKATION, Lenzkirch

The Lenzkirch Clock Factory, or Aktiengessellschaft für Uhrenfabrikation, was founded by Eduard Hauser who was born August 21, 1825, the son of a teacher. Eduard Hauser first gained experience making music boxes in 1840 with Johann George Schöpperle. Hauser learned many skills related to this apprenticeship, such as metal working, precision mechanical work and the design of musical instruments. During this time he also gained a knowledge of musical composing.

Eduard Hauser, perhaps becoming restless and wanting to extend his mechanical know-how, felt that clockmaking might be a better business in which to be involved. He left the small music box business and studied clockmaking in England, France, and Switzerland. Hauser felt that if he could put the precision factory methods of the French, English, and Swiss to work for him in the Black Forest, it would certainly make the area competitive with the rest of the precision clockmaking countries. Up to this point in time, Black Forest clocks, such as cuckoos, did not have a good technical basis or reliable reputation for precision.

Hauser first worked with Ignaz Shöpperle, opening a small machine shop where clock parts were made for other clockmakers. Hauser was trying to introduce a method whereby precision parts could be passed to an assembly line and the movements would be put together versus the old

The 8-day movement and 2 spiral gongs. Movement is stamped with "W & H in SCH, Germany." (Antique Clocks Publishing Archive).

Winterhalder & Hofmeier bracket clock with mahogany case and brass dial with silvered metal chapter ring. Adjustment to chime or silent on the right with time adjustment on the left. Sold by Mappin and Webb in Manchester, England. (Antique Clocks Publishing Archive).

W & H Baroque style bracket clock with original bracket, or shelf. Brass dial with silvered chapter ring. 8-day movement with strike on 8 bells or you can adjust to strike on 4 spiral gongs at the hour. Above the dial are adjustment for time, chime or silent, and bell or gong strike. (Antique Clocks Publishing Archive).

The Winterhalder & Hofmeier 8-day movement, quarter strike with strike on 2 spiral gongs at the quarter and strike on 1 gong at the hour. Jeweled platform escapement. (Antique Clocks Publishing Archive).

Catalog page from Winterhalder & Hofmeier, about 1925. Depicts two hall or grandfather clocks. (Antique Clocks Publishing Archive).

Catalog page from W&H, about 1925. Depicts 3 mantel clocks. (Antique Clocks Publishing Archive).

Catalog page from Winterhalder & Hofmeier, about 1925. Depicts English style bracket clocks. (Antique Clocks Publishing Archive).

method of the clockmaker making every part by hand, assembling the clock, and praying that it would turn out okay and run correctly. Both partners received technical aid from Robert Gerwig, the director of the newly established clockmaking school in Furtwangen, who also believed in this new method of clock assembly.

Working in Shöpperle's organ factory, Hauser supervised his employees and produced the precision clock parts using machinery which along with fixtures and raw materials, was bought new, draining the accounts of both partners. There was a real need for more cash flow to market the movements they manufactured.

On August 31, 1851 Eduard Hauser joined forces with Franz Josef Faller (1820-1887), Nikolas Rogy, Johann Nikolaus Tritscheller (1825-1867), Paul Tritscheller (1822-1892), and Joseph Wiest, to form Aktiengesellschaft für Uhrenfabrikation. In 1856 Albert Tritscheller (1833-1889) also joined the others and spent time learning new manufacturing methods in other countries.

In 1857 Franz Josef Faller, Paul and Nikolaus Tritscheller, and Eduard Hauser were guiding the clock factory. Eduard Hauser designed machinery far advanced for this time period, and made precision clock parts from his automatic machinery mills. Under his leadership the Lenzkirch clock factory developed and gained a solid reputation for

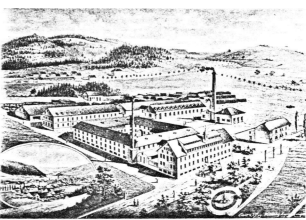

Edward Hauser, the founder of the Lenzkirch Clock Factory. Born August 21, 1825, died July 22, 1900. (Antique Clocks Publishing Archive).

The Lenzkirch clock factory in the year 1928, from a Lenzkirch catalog. (Antique Clocks Publishing Archive).

EDUARD HAUSER

making excellent and reliable timekeepers. It gained a reputation that no other German factory had acquired so far.

Eduard Hauser wanted his sons to continue their educations and hoped that some day they would take over the reigns of the factory. His oldest son was an excellent architect and the Lenzkirch clock cases reflected his influence for about thirty years. His son Karl August was apprenticed under Jess Hans Martin (1826-1892), master clock and watchmaker in Freiburg/Breisgan and teacher at the Badenia Trade School for clock and watchmakers in Furtwangen. This prepared Karl to take over and supervise the tool and die making operation.

Paul Emil Hauser, the youngest son of Eduard, helped in the development and assembly of machinery. His training as a tool and die maker helped prepare him for the aforementioned task.

All of the people mentioned previously helped the Lenzkirch clock factory achieve a reputation

170 Black Forest Clocks

Black Forest alarm clock in a very small but ornate walnut case, measuring just under 12 inches tall. Made by the Lenzkirch factory about 1895. The stamp of the factory is on the dial. (Deutsches Uhrenmuseum, Furtwangen/Schwarzwald, photo Callwey Verlag, München, Germany).

for making good quality, reliable clocks. The regulator clocks had a worldwide reputation for their quality and reliability. These came into full production about 1860. Lenzkirch was awarded gold medals in the industrial exhibitions in 1860 in Villingen and 1861 in Karlsruhe. Weight and spring driven clocks became very popular, the latter having many different style from which to choose. Lenzkirch made their own springs for their clock movements, and also sold them to other clock factories.

Franz Joseph Faller helped to market the Lenzkirch clocks by devising a sample catalog of inventory that was on hand or could be manufactured. There was a wide selection of models that were produced. The customers could order from the catalog; once the order was received in their office the Lenzkirch factory would then fill the order.

Faller also worked with Paul Tritscheller to connect the city of Lenzkirch with the existing railroad system, but in the meantime horse-drawn wagons carried clocks over mountain roads and stormy conditions to the railroad stations at Todtnau and Hammereisenbach. Winter months seemed to hinder all clock shipments. By railroad, clocks were shipped to various parts of Europe for further shipment to England, the U.S. and other parts of the world. In 1887, the railroad system was connected from Freiburg to Neustadt.

1887 was also the year Faller died, and the Black Forest had lost one of its leaders in the clockmaking

Freeswinger made by the Lenzkirch Clock Factory. About 60 pieces of brass are applied to the ornate walnut case. 14-day running duration. Circa 1890.

Lenzkirch wall clock with brass decoration. The case is very ornate with columns, finials, railing, carving, etc. The dial is metal and silvered. 14-day running duration, brass plate movement with the stamp of the Lenzkirch factory. The case is made of walnut. 50 pieces of brass applied to the case which measures 43 inches in length. Circa 1880.

The brass pendulum bob of one of the most ornate Lenzkirch clocks ever made. (Eugene Kramer collection)

Freeswinger made by the Lenzkirch Clock Factory. Walnut case with 50 pieces of brass decor is 37 inches in length. 14-day movement. Circa 1890.

Small desk clock made by the Lenzkirch clock factory. The boy swings on the pendulum. Case is walnut with brass decoration and is 12 inches tall and 4 inches wide. Time only. Circa 1880. (Cleveland, 1986 NAWCC National Convention)

Very large model number 93 Regulator made by the Lenzkirch Clock Factory. Case is 6½ feet in length. Sweep second hand. 8-day running duration.

industry. He was the one who had initially decided that the manufacture of Black Forest clocks should be done on a large, production-type method in order to become competitive throughout the world. Faller helped to change the production methods of the Black Forest clockmakers. He also helped the city of Lenzkirch by establishing a watch and clockmaking school which made it possible for other clockmakers to develop large scale industrialization. He still supported the small cottage clockmakers and believed they should also survive in the Black Forest, producing clocks in the same way they always had. He did not want to see them merge with other large clock factories.

The Lenzkirch clock factory employed thousands of workers over an almost 80 year period, at one time employing a work force of 600 workers. During this entire time period, they made every effort to continue to produce quality clocks and always kept the Lenzkirch Clock Factory's reputation intact. Because of this, business flourished most of the time, the exception being in times of Depression in 1866, 1870, 1876 and 1892.

Edward Hauser retired March 1, 1899. In 1900 the company employed 480 workers, the largest in the Badenia Black Forest. Eventually two of Hauser's sons went to work for a competitor—The Hamburg American Clock Co. in Schramberg. The competition in this clockmaking area of the Black Forest eventually caused the Lenzkirch factory to merge with the Gebrüdder Junghans firm of Schramberg. In the middle of the Great Depression in 1932 the buildings were sold by the Junghans Brothers, who operated it as a satellite factory for about four years, and then the famous factory of the Lenzkirch Clock Company was gone.[3] The clocks, however, live on and have today become very popular among clock collectors throughout the world. Their products helped to establish the Black Forest clocks' worldwide reputation for quality.

It is the author's opinion that the Lenzkirch factory was the best industrialized factory in the Black Forest. No other factory decorated as many of their cases in the way they did. Brass decoration and walnut and oak were used on many of their cases. Richly decorated and high grade movements make them accurate and attractive to timekeepers today. The LFS and Winterhalder/Hofmeier clock cases came close to Lenzkirch standards, but Lenzkirch, it seems, was the best. Some Lenzkirch clock cases are stamped with the factory name on the back of the case. In the late 1870s the firm could not stamp their movements with the trademark when exporting to certain countries. That is why you will find unmarked movements on many Lenzkirch clocks.

Standard size Lenzkirch spring driven model with half hour strike. Walnut case with 2 carved figures is 33 inches long. Top ornaments are not original. The original top ornament was probably much more ornate and should have matched the rest of the case carving. 14-day movement. Circa 1870.

Regulator made by the Lenzkirch Clock Factory with half hour strike. Walnut case with black trim, original to the finial. Door has a key lock. 8-day movement. Circa 1880.

174 Black Forest Clocks

The Lenzkirch trademark used around 1880. (Photo by Keith Lee)

Movement of the Lenzkirch RA. The serial number on the movement is also stamped on the metal backplate behind the porcelain dial.

Spring driven clock made by the Lenzkirch Clock Factory. Early RA style, serial number in the 30,000 series. Walnut veneer case with black trim. 14-day movement. Circa 1880.

KARL JOSEPH DOLD SOHNE UHRENFABRIK, Schönwald

In 1842 Joseph Dold started with a small clock shop and made the popular types of clocks, while concentrating on cuckoo clocks. In 1884 his son Karl Joseph Dold took over and expanded the company to a large mechanical mass producing factory by 1894. They made wall, floorstanding, and alarm clocks.

After his retirement in 1911, Karl's sons Alfred and Adolf Dold ran the company. In 1925 they employed approximately one hundred people, including the people who worked in their own homes. In 1961 the company could not keep up with their competitors and closed.[4]

WEHRLE UHRENFABRIK GmbH, Schönwald

In 1815 Andreas Hilser started this company, making his own clocks and traveling to sell them. His son Raimund Hilser took over after his death, and later Raimund's son-in-law Carl Wehrle also took over and helped run the factory. He expanded and in 1890 started making alarm clocks patterned after the American system. His son Carl Joseph Wehrle (1880-1968) took over in 1910 after his father died in an accident. After Carl Joseph's death in 1968, his sons Karl Raimund and Frans Wehrle took over.

The company Wehrle Uhrenfabrik Gmbh Schönwald has 250 employees today. It is the most famous alarm clock factory in the Black Forest, exporting its product around the world. It has a satellite factory in Fischerbach bei Haslach in Kinzigtal.[5]

Footnotes

[1] Gerd Bender, *Die Uhrmacher des hohen Schwarwaldes und ihre Werke*, pp. 150-151, Verlag Müller, Villingen, 1978.

[2] Karl Kochmann, *The Lenzkirch-Winterhalder & Hofmeier Clocks*, pp. 85-92, Antique Clocks Publishing, Concord, 1984.

[3] Karl Kochmann, *The Lenzkirch-Winterhalder & Hofmeier Clocks*, pp. 5-28, Antique Clocks Publishing, Concord, 1984.

[4] Gerd Bender, *Die Uhrmacher des hohen Schwarzwaldes und ihre Werke*, p. 142, Verlag Müller, Villingen, 1978.

[5] Gerd Bender, *Die Uhrmacher des hohen Schwarzwaldes und ihre Werke*, p. 141-142, Verlag Müller, Villingen, 1978.

Large Lenzkirch quarter-hour strike, 14-day wall clock with strike on 2 gongs for the quarter and 1 for the hour. 85 pieces of brass decoration applied to the walnut case which is 49 inches long. Circa 1890.

Lenzkirch 14-day time-and-strike mantel clock with yellowed porcelain dial and oak case with brass decoration, 20 inches tall. The top piece is all brass. Circa 1895.

A Lenzkirch (Aktien Gesellschaft für Uhrenfabrikation) mantel clock. The case is made of oak. Metal silvered dial. 14-day movement. Circa 1900.

Lenzkirch freeswinger in a walnut case, veneered with a few pieces of brass decoration on the top of the case. 14-day time-and-strike movement. Circa 1890. Case measures 38 inches long.

The Clockmakers of Neustadt, Lenzkirch, and Schonwald 177

Lenzkirch mantel clock patterned after the English bracket clock style. Etched metal dial. The case measures 12 inches tall with brass decoration. 8-day movement with balance escapement mounted on the backplate which strikes on a spiral gong at the half and full hour.

Louie XV style, time only mantel clock made by Lenzkirch. The case measures 12 inches tall with brass decoration.

A small Lenzkirch wooden case alarm clock with brass decoration. Enamel dial. Serial #171103 stamped on the movement.

178 Black Forest Clocks

Lenzkirch miniature wall clock 21 inches in length. Black case with black porcelain dial. Brass decoration. 14-day time-and-strike movement. RA pendulum. Circa 1865.

Quarter strike Lenzkirch wall clock that measures 42 inches long. Enamel dial. Circa 1900.

Lenzkirch mantel clock that is 20 inches tall with brass decoration. Time-and-strike 14-day movement. Missing top finial.

Oak wall clock by the Lenzkirch factory that measures 35 inches long with 46 pieces of brass on the case. Brass mermaid on top. Silvered dial. Circa 1885.

Lenzkirch wall clock with quarter hour strike. Case is dark and measures 36 inches long. Enamel dial. Circa 1900.

Gallery style Lenzkirch wall clock. Pendulum is not visible. The case measures 33 inches in length. Brass decoration on the oak case. Circa 1890.

Lenzkirch mantel clock with the original shelf. Case is 21 inches tall, the shelf is 8½ inches. Both have brass decoration. Round French style time-and-strike movement which strikes on a bell.

Baroque style Lenzkirch wall clock, the pendulum is not visible. Case is 31 inches long.

Small Lenzkirch mantel in a marble case. 14-day movement, time-and-strike. Circa 1900.

Lenzkirch mantel clock with the original shelf. Case is 13½ inches tall. Shelf is 7 inches tall. The top is missing from the clock. 14-day time-and-strike movement.

Lenzkirch 2 weight Regulator. Brass dial, weights and pendulum bob. Case measures 49 inches long. 8-day movement. Circa 1885.

182 Black Forest Clocks

Lenzkirch freeswinger in an unusual case. Time-and-strike 14-day movement. Circa 1890.

Lenzkirch gallery style wall clock, missing the top decoration. Circa 1890.

Lenzkirch RA wall clock with 14-day time-and-strike movement. Circa 1890.

Small Lenzkirch mantel clock with alarm, unrestored in a walnut case. 16 pieces of brass decoration. Circa 1895.

Lenzkirch freeswinger with 14-day time-and-strike movement in an oak case. 42 pieces of brass decoration on the case. Circa 1885.

Lenzkirch time only, in a brass case with silvered dial. Circa 1900.

184   Black Forest Clocks

Lenzkirch "balloon" style case clock with enamel dial. The top finial is broken. Case is about 12 inches tall.

Lenzkirch mantel clock in a brass case with beveled glass. Circa 1900.

Lenzkirch 14-day time-and-strike wall clock, open well type with 20 pieces of brass decoration applied to the walnut case that is 38 inches long. Circa 1885.

An elaborately made Lenzkirch open well clock adorned with the figures of angels and various masks made of brass. Silvered dial. Circa 1880. Case measures 40 inches in length.

A nice Lenzkirch mantel clock that is 24 inches tall with brass decoration. Round French style movement. Circa 1890.

Small 8½ inch high mantel clock with balance type movement stamped Lenzkirch.

186 Black Forest Clocks

Small walnut Lenzkirch alarm clock with the trademark on the dial. Walnut case measures 9½ inches.

Number 318 Regulator, time only, 8-day movement by Lenzkirch measuring 54 inches in length. The dial reads "C.F. Wolf Hannover, Germany" (the retailer).

Small Lenzkirch walnut case, time only mantel clock with etched dial and brass decoration. Circa 1900.

A group of Lenzkirch alarm clocks. The left and right clocks are encased in wood, and the one in the middle is made of brass.

Porcelain-cased Lenzkirch time-and-strike mantel clock. The porcelain is stamped "PLAUE," and was made by a factory in Thuringia, Germany, circa 1880-1906. The factory was C.G. Schierholz & Son.

Lenzkirch freeswinger with quarter-strike movement on 2 gongs, 1 on the hour. Brass decoration applied to the walnut case that measures 44½ inches tall. Circa 1890.

Small Lenzkirch mantel clock made from walnut. Case measures 18 inches tall with brass decoration. The swing acts as a pendulum. Circa 1890.

Small Lenzkirch time-and-strike mantel clock. Case measures 10½ inches tall. Circa 1910.

Lenzkirch clock 13½ inches tall. Round French style movement. Porcelain and brass case. Time only movement.

Lenzkirch mantel clock which measures 24 inches tall. Quarter strike on two gongs, one on the hour. Walnut case with brass decoration. Circa 1890.

Lenzkirch mantel clock, black case with painted decoration measures 19½ inches tall. Porcelain dial. Circa 1900.

190 Black Forest Clocks

Lenzkirch model number 321 Vienna style 2 weight Regulator. Time-and-strike movement. Carved birds and deer head at the top. Bird, rabbit, rifle and hunter's pouch on the door. The veneer on the case, which measures 48 inches in length, is burled. Circa 1870.

Lenzkirch quarter strike movement mantel clock that measures 16½ inches tall. Slow/fast adjustment above the dial. Case is burled and carved at the top with brass decoration.

Oak mantel clock with brass decoration made by Lenzkirch which measures 21½ inches tall with the original shelf that measures 8½ inches. Time-and-strike 14-day movement.

The Clockmakers of Neustadt, Lenzkirch, and Schonwald 191

Lenzkirch mantel clock, veneer on the case is burled and measures 21 inches tall with brass decoration. Enamel dial. Circa 1880.

Lenzkirch wall clock with 14-day time-and-strike movement. The case measures 44 inches in length. Circa 1895.

Lenzkirch time-and-strike mantel clock that measures 22 inches tall. Silvered dial. Circa 1890.

Lenzkirch wall clock with RA pendulum and outside escapement. Time-and-strike 14-day movement. Case measures 37 inches. Circa 1880.

Number 365 Lenzkirch 2 weight Regulator with 8-day movement. Circa 1875.

One weight Lenzkirch Regulator. Case is burled and measures 53 inches long. Enamel dial. Circa 1880.

Lenzkirch model number 83 weight driven clock, time-and-strike, 8-day movement. Circa 1870. Case measures 48 inches in length.

Lenzkirch wall clock with hour and ½ hour strike. The oak case is 40 inches long with copper and pewter dial and pendulum. Circa 1890.

Lenzkirch keyhole or peanut style. Case is black and measures 28 inches long. Ornate RA pendulum. Circa 1870.

194 Black Forest Clocks

A Lenzkirch mantel clock with original bracket, all made from oak. Metal, silvered dial which is etched. Circa 1890. This clock was purchased in 1971 in Germany by a serviceman who brought it back to the USA for his collection.

Lenzkirch time only wall clock which measures 23 x 17 inches, the case is made of walnut. Brass decoration. Dial is 7 inches in diameter.

A nice Lenzkirch mantel clock that measures 16 inches tall with quarter hour strike on 2 gongs. The wood veneer on the case is burled and has brass decoration. Circa 1890.

Lenzkirch mantel clock with brass decoration. Case is 21 inches high and made of mahogany. Sunburst pendulum. Circa 1890.

Large wall clock which measures 51 x 20 inches, made by the Lenzkirch factory. Time-and-strike 14-day movement. Circa 1890.

Lenzkirch gallery clock, time only. The case measures 14 inches long.

Brass case mantel clock with marble base and back made by Lenzkirch. The case is 15½ inches tall and the dial is 3 inches in diameter. 14-day time-and-strike round plate movement.

Small Lenzkirch bracket clock that measures 11 x 6 inches with a 5 inch bracket. Silvered dial. Case veneer is burled with applied brass decoration.

A nice oak case Lenzkirch mantel clock that measures 18 x 13 inches with quarter hour strike on 2 wire gongs. Circa 1895.

Another small Lenzkirch bracket clock that measures 12 x 8 inches with a bracket that measures 6 inches. The enamel dial is 4 inches in diameter. Burled wood with brass decoration. 14-day time-and-strike movement.

Lenzkirch mantel clock. The case is 14 inches tall and the veneer is burled. 14-day time-and-strike movement. Circa 1890.

CHAPTER 9

Trumpeter Clocks

One type of Black Forest clock that is unusual and difficult to find is the trumpeter clock. The musical trumpeter clock originated from the area of Furtwangen. Legend has it that a local artist who painted for royalty, Johann Baptist Kirner, suggested the idea of a trumpeter clock in 1857. Drawings were made depicting a man in a doorway holding a trumpet. The strike portion of the clock would sound like a trumpet call, not a bell or a gong. Jacob Bäuerle was the first clockmaker to experiment with this type of clock. It also appears that the poem written by Viktor von Scheffel called "Trumpeter von Säckingen" contributed to the trumpeter clock idea.

JACOB BAUERLE

Jacob Bäuerle began with a basic cuckoo type movement, but instead of a cuckoo pipes, he used two pipes which copied the sound of a trumpet. He had problems with this and found that it did not work well with the basic cuckoo movement. The movement he made could not make a realistic sound of a trumpet, so Bäuerle built a clock that copied the sound of a train conductor's call. A train conductor came out of the door in this type of clock. The cost of these clocks was high and the average person could not afford them, and they certainly could not compete with the less expensive cuckoo clocks being made. Bäuerle kept experimenting and made very elaborate clocks which included many types of trumpeter, soldier, and postman clocks that sounded two or three notes at the hour.

In 1858 Bäuerle won an award at the clock industry exhibit of Villingen for the so-called trumpeter clock. His clocks were very popular with the people, but were also very expensive. Bäuerle's son Karl also made trumpeter clocks with great success in the mid-1870s, making a variety of both large and small clocks with varied musical tunes. He kept improving the trumpeter clocks and is credited with making the first flute clock in this period. While the technology existed for eight-day models, they were quite expensive, and Bäuerle's clocks were made to run for 24 hours. The making of these types of clocks with music was so very complicated that only two shops made them: Jacob Bäuerle and Emilian Wehrle.[1]

EMILIAN WEHRLE, Furtwangen-Schönenbach

Emilian Wehrle, was born in 1832 in Schönenbach. Emilian's father, Michael, was a small farmer and clockmaker, but not very well known. Emilian learned the clockmaking trade from his father while still in school. When Emilian was called to serve in the army, he served extra time to earn more money so he could start his own business after his discharge. In his handwritten diary he explains the beginning of his factory in 1857. He worked day and night, and by 1858 stated that he was doing well.

Around this same time he was called back into the military, which was a setback for his business. But by 1859-60 it was running smoothly again. By 1859 he was worth 523 marks and 30 pfennig. One year later his worth had doubled to 1001 marks and 60 pfennig. His clocks were well received at the clock industry show in Karlsruhe, Germany in 1861. In 1864 he produced 160 clocks. One year later the factory production rose to 190 clocks, including both trumpeter and cuckoo clocks.

In 1860 Emilian married Norma Wehrle, the oldest daughter of his neighbor F.X. Wehrle who was a music box manufacturer. This became a business connection and a new company was formed and called Emilian Wehrle and Co., producing mainly trumpeter clocks. However, Emilian was always experimenting and eventually produced flute clocks, rooster call clocks and, in 1875, a clock with a mechanical singing bird.

Trumpeter Clocks

The front side of a wood plate movement with brass gears and steel arbors, and the windchest at the top of the movement that sounds on the hour on one pipe. The call is similar to that of a Rooster and was probably made by Jacob Bäuerle. Jacob is said to have been the first clockmaker to experiment with clocks that would strike like a trumpet call, not a bell or a gong.

Carved Jacob Bäuerle flute clock with 10 wooden pipes. The case measures 42" x 30" x 15". The movement is mounted above, and the windchest and pipes are below, unusual as they are commonly placed on top of the movement. The wood cylinder is 4¼" diameter by 4" long and plays three different tunes.

The backside of the early movement with wheel for rooster call.

Sideview showing the pipes, lifting wires and wood, brass pinned music cylinder and windchest underneath.

Above, the brass movement; below, the wood cylinder, wood pipes and wind chest.

The flute player and intricate bottom clock showing the detailed carving.

The wood music wheel and apparatus connected to the brass movement above.

Emilian Wehrle (1832-1896). Operated a trumpeter clock factory in Furtwangen-Schönenbach/Moos. Also made rooster call and singing bird clocks. (Antique Clocks Publishing Archive)

Wehrle won many awards for various types of clocks through the years but the production of trumpeter clocks stopped in 1896 when Emilian died. Emilian's brother-in-law, Julian Wehrle, who had joined the firm earlier, continued with the factory after Emilian died. But financial trouble developed and he finally turned the factory towards specializing in fine mechanical parts.[2]

The trumpeter movements had brass horns on both sides which were connected to a large windchest located at the top of the movement. Inside the windchest were small brass organ reeds. These reeds, along with the proper size metal pipe attached to the windchest, produced the sound of various notes on a musical scale when in tune with the musical wheel located on the backside of the movement. The musical train on the shelf model trumpeters had to be wound daily, although the time and strike portion lasted eight days. The author is interested in these types of clocks because of the musical tunes that play on the hour. They sound similar to an accordion, and, when tuned and repaired properly, they can fill a house with a most pleasant sounding tune. Wehrle also made weight driven models that hung on the wall but would run for only one day. If the clock had two tunes, one only had to shift the music wheel to select the tune that was desired. The author has seen trumpeter and flute clocks, made by Wehrle, with as many as eleven pipes. It is possible that some were made with even more.

On a flute clock, the difference in the sound played was that the pipes that produced the musical sound were made of wood and had no organ reeds in the windchest. When the music played the clock sounded like a flute player blowing his flute.

A person need only hear one of these musical type clocks in proper playing condition to understand the author's interest in all Wehrle clocks. It is the author's opinion that the trumpeter clock plays and sounds better than the flute clock. Also, the carved cases made for the Wehrle factory were well done, but the architectural cases seemed to be made in a more solid, stable manner. Perhaps this was because Mr. Tritscheller designed them well; they seemed to have withstood time better than the carved examples.

Footnotes
[1] Gerd Bender, *Die Uhrmacher des hohen Schwarzwaldes und ihre Werke*, pp. 277-282, Verlag Müller, Villingen, 1979 revised.
[2] *Die Uhrmacher des hohen Schwarzwaldes und ihre Werke*, pp. 282-294, Verlag Müller, Villingen, 1979 revised.

The Emilian Wehrle Trumpeter Clock Factory in Furtwangen, Schönenbach. This lithograph was used as a letterhead around 1870. (Antique Clocks Publishing Archive)

Thirty-hour wall trumpeter, weight driven with eight metal horns and two musical tunes.

The musical movement with eight wooden pipes and original pendulum. The Wehrle label is usually found on the windchest on top of the movement.

An eight-day running flute clock made by E. Wehrle with the hunters motif: the deer head, rabbit, fox, and dog at the bottom of the 44 inch tall case. The hands are carved from bone and the numerals are made of brass. This clock is in very original condition.

The painted flute player which appears at the hour.

The label on the back of the flute clock case.

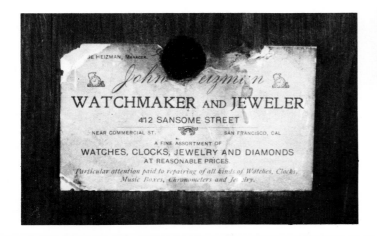

Trumpeter Clocks

Advertisement for an E. Wehrle, Number 13 trumpeter or flute shelf clock. (Dr. Wilhelm Schneider, Regendorf, Germany).

The front side of the Wehrle flute movement and windchest. Note the large spring barrel for the musical train of the clock which must be wound daily.

The backside of the Wehrle movement shows the music wheel and a label on the windchest which at one time listed the songs played, however the writing is faded and the label is ripped so the tunes can no longer be read.

Number 13 shelf flute clock with eight wooden pipes made by E. Wehrle in Furtwangen. The carving is from walnut. Bone hands and bone numbers. The case measures 44 inches tall. This model plays one of two tunes which may be selected by shifting a lever that moves the music wheel attached to the back of the brass movement. The flute player appears at the hour and is made of a plaster-like material and painted.

204 Black Forest Clocks

An example of a finely carved, unrestored, Black Forest trumpeter clock. The movement is a four-horn type with a running duration of thirty hours. It is shown in an unrestored condition only because of the fine carving.

The full relief carved eagle at the top of the case.

The three birds at the bottom of the case.

Trumpeter Clocks 205

The frontside of the windchest for the four-horn trumpeter clock made by Emilian Wehrle.

The backside of the windchest with the tune written in German.

The windchest taken apart to show the four wind tunnels and four organ reeds. The musical note is written on the side of the brass reed.

The bottom of the windchest.

206   Black Forest Clocks

Trumpeter clock made by Emilian Wehrle, Furtwangen, Germany, about 1880. The highly carved case is about 48 inches in length. Plays two tunes on eight brass pipes. Walnut case.

The deer at the top of the Wehrle trumpeter, beautifully carved. Original antlers. On many clocks the antlers are missing or broken, or are in many pieces. They must be reglued, doweled, and re-carved to add the broken tips.

The movement made by Emilian Wehrle. Note the large bellow or windchest and the music wheel. The wheel shifts for a second tune, if desired.

The two trumpeter figures are activated when the tune plays. Molded from a plaster-like substance.

The birds nesting at the bottom of the Wehrle trumpeter clock.

The dial and the key and crank to wind the Wehrle movement. Bone hands. Note the fine detailed carving around the dial.

A flute clock or Flotenuhr made by Emilian Wehrle and Company about 1880. Stag or deer head located at the top of the case with leaf and vine carving. Intricate bone hands. The clock strikes at the indicated hour strike and the doors at the bottom open revealing a flute player. The clock then plays a musical tune on eight wooden pipes. A fine example of a Wehrle clock.

208 Black Forest Clocks

Wall trumpeter clock with eight metal horns made by E. Wehrle and Co. about 1880. Eagle and deer motif, exquisitely carved. Bone numbers and hands. Trumpeter doors are not original. The music in this clock is very loud, but also very pleasant sounding.

The eagle at the top of the trumpeter clock.

The deer at the bottom of the walnut case.

Trumpeter Clocks

The sitting deer at the bottom of the case.

The trumpeter figure.

Floorstanding trumpeter clock made by E. Wehrle about 1885. Nine brass horns. Plays two tunes, one at the hour. Metal dial. The floorstanding style is seldom seen in a trumpeter; more shelf and wall models were made.

The two painted trumpeter men and a closeup of the elaborate moldings used on the case.

The movement, windchest, pipes, and music wheel on the right.

Trumpeter clock made by E. Wehrle and Co., circa 1890. Spring driven eight-day movement although the music train will run for only a day. Four brass pipes. Painted figure at the bottom behind the door. Metal dial with intricate metal hands.

Trumpeter Clocks 211

The back of the clock showing the movement, windchest, pipes, and music wheel.

Trumpeter clock made by Emilian Wehrle & Co. about 1885. Gothic style case. Plays two tunes, one on the hour. The movement runs for eight days.

The windchest with the EWC Fabrik label.

The two trumpeter figures which appear at the hour.

Nicely carved wall trumpeter by E. Wehrle, with thirty-hour weight driven movement which has eight brass horns for the music that plays on the hour. Enamel numerals; the minute hand is a replacement. (Antique Clocks Publishing Archive).

Trumpeter clock by E. Wehrle. Wall model with thirty-hour movement. Hunter motif. Nine horns with two tunes. Original pewter hands. Circa 1890.

The weight driven movement with wood music wheel and windchest with eight brass pipes. The label would have had the song written on it but it is long faded. (Antique Clocks Publishing Archive).

Trumpeter Clocks

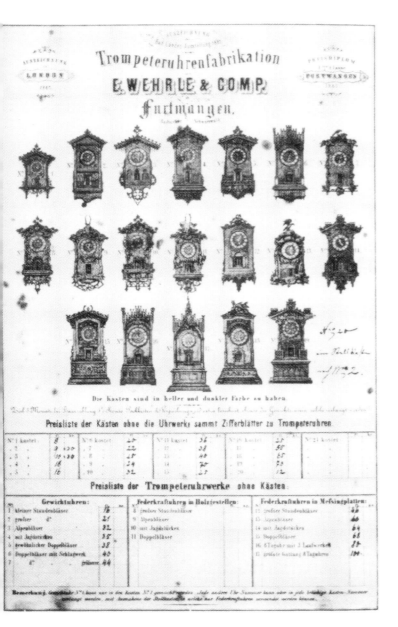

An E. Wehrle advertisement taken from his 1866 catalog. Shows various models of shelf and weight driven trumpeter clocks and price list. (Antique Clocks Publishing Archive).

Number 21 trumpeter or flute clock by E. Wehrle from his catalog. Height about 37⅜ inches. Weight driven with strike on a spiral gong, it could be ordered with brass pipes and organ reeds for the trumpeter or wooden pipes for the flute clock. (Dr. Wilhelm Schneider, Regendorf, Germany).

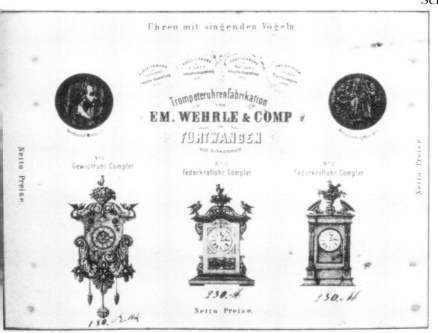

An E. Wehrle 1866 catalog advertisement for weight and spring driven singing bird clocks. (Antique Clocks Publishing Archive).

214 Black Forest Clocks

A special trumpeter clock made by E. Wehrle about 1890. This clock is fitted with an eight-day, eleven horn, two tune musical movement. Walnut case with metal, silvered, etched dial.

The two painted trumpeter figures.

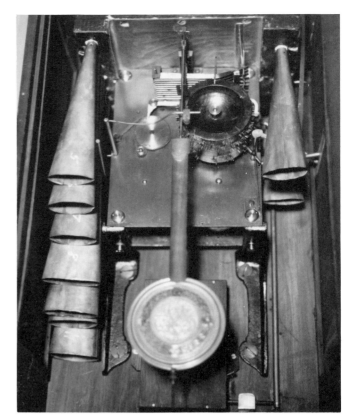

The eleven horn brass pipe trumpeter movement which shows the brass pinned wood music wheel and brass lifters causing the reeds to sound off through the brass pipes.

View of the embossed pendulum.

A castle style trumpeter clock made by the Wehrle factory. The case measures 35 inches tall by 24 inches wide. The movement is eight-day, strikes on a wire gong and plays a tune at the hour on eight brass pipes. The doors to hide the trumpeter figures are missing. (Photo by Doug Barr, Berea, Ohio)

A number 15 "Doppelbläser" seven horn trumpeter clock made by Emilian Wehrle. The same clock is pictured in the lower left of the 1866 catalog page. The hands should have been much more intricately carved for this clock. Eight-day movement for the time and the musical train of gears must be wound daily.

The side door of the "Doppelbläser" trumpeter.

The two painted trumpeter figures.

216 Black Forest Clocks

An advertisement from an E. Wehrle catalog page showing the number 15 Castle style case flute clock, or Flötenspieler. This model would run for eight days by a heavy brass plate, brass gear movement, and has two musical tunes. The clock case is 37¾ inches tall. The advertisement does not specify how many wooden pipes, however it could have had from four pipes up to possibly eight pipes. (This ad was provided by Dr. Wilhelm Schneider of Regendorf, Germany.)

An advertisement for the Trumpeter from Säckingen clock, No. 33 Scheffel-Uhr. (Dr. Wilhelm Schneider, Regendorf, Germany)

Trumpeter Clocks 217

The brass movement and rooster call apparatus, patent number 32141. Emilian Wehrle was certainly an imaginative inventor of his time. The rooster clock, or Hahnen-uhr, was patented by Emilian Wehrle in Furtwangen in 1884. This simple apparatus makes the sound of the rooster call without using pipes or horns. A newspaper article dated October 9, 1885 from Frieburg, Black Forest, congratulates Wehrle on his efforts, talent, and fine workmanship.

This 19 inch tall clock chimes with the call of the rooster. The rooster at the top moves with the call. The clock is signed E. Wehrle and Co. behind the wood bezel ring around the dial. You have to remove it to find the signature. So before you think your clock is unsigned, look a little deeper. You might be surprised by what you find.

The metal rooster which moves at the hour call.

218 Black Forest Clocks

Large, heavily carved trumpeter clock by E. Wehrle. Eight horn musical movement, eight-day with two tunes. Bone hands. The author has seen one other clock like this, but the example shown is in mint condition. Circa 1885.

The carving on the right side, bird and deer.

View of the dial and large bird at the top of the case.

Trumpeter Clocks 219

Castle style four-horn shelf trumpeter clock with bracket. Eight-day time strike and music. The trumpeter figure sits on a horse. Circa 1895.

The frontside of a thirty-hour weight driven eight horn trumpeter movement by E. Wehrle. Circa 1880.

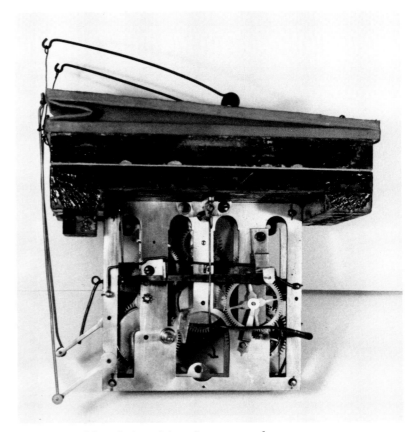

The backside of the thirty-hour cast brass trumpeter movement with wind chest. The music wheel is removed.

Four-horn trumpeter movement from E. Wehrle. Time and music only. There is no gong strike, so the movement will run for thirty hours.

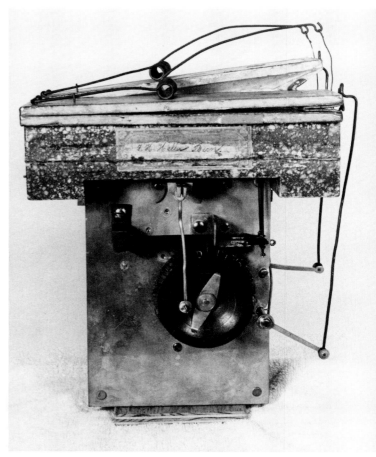

The rear of the four-horn movement with wooden, brass pinned music wheel and label listing the single song on the rear of the windchest.

A Wehrle windchest for a seven-horn trumpeter clock.

Trumpeter Clocks

The windchest taken apart in two sections revealing the seven wind chambers and seven brass organ reeds. The tongue of the reed is very fragile and is critical to the sound forced through the pipe. If damaged or cracked, the note may sound different than it should. If you place the wrong pipe in the hole it will also change the sound of the note. Also, if the pipes to your trumpeter movement are missing, the tune played without the pipes will sound different than if the pipes were attached to the windchest. The author advises that only knowledgeable clock restorers who have previously worked on musical mechanisms and are also knowledgeable in music be hired to restore these timepieces.

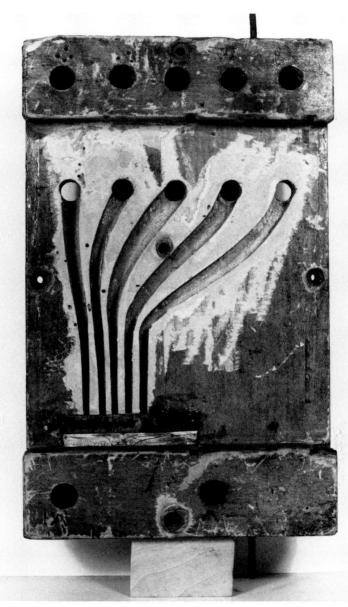

The underside of the windchest revealing the wind tunnels. Paper or leather normally covers these tunnels. However, as in the case of many clocks, it was rotted away and needed replacement.

222 Black Forest Clocks

Nr. 1.
Höhe 96 Centimeter
Federkraft 8 Taguhr
mit mechanischem Singvogel.
Ziffer und Zeiger vernickelt.
Complet.

Nr. 2.
Höhe 70 Centimeter
Federkraft 8 Taguhr
mit mechanischem Singvogel.
Ziffer und Zeiger vernickelt.

Nr. 3.
Höhe 70 Centimeter
Gewichtuhr 24 Stund gehend mit
mechanischem Singvogel.
Ziffer und Zeiger vernickelt.
Complet.

Three singing bird clocks from an Emilian Wehrle catalog. Various size. The two shelf models are eight-day with nickel numerals and hands. Number 3 is weight driven, one-day with nickel numerals and hands. Circa 1885. (Dr. Wilhelm Schneider, Regendorf, Germany).

Eight horn, spring driven wall model trumpeter by E. Wehrle with the hunters motif. Movement is eight-day. Clock case has the original carved bottom side pieces, which are often missing on the wall models. The hands would have been more intricately carved from bone. Circa 1875.

Trumpeter Clocks

Number 33 Scheffel-Uhr, trumpeter from Säckingen (Trompeter von Säckingen) by Emilian Wehrle, made about 1890. Eight-day musical movement has two solid brass plates with brass gears and steel pinions with nine metal pipes and two musical tunes. After the trumpeter tune plays, a small lever is tripped and one of four tunes play from a music box which is located in the base of the clock. The clock case (walnut) measures 60 inches in length and the bracket (also walnut) that it sits on is 24 inches long. The top console has 206 various sized pieces of brass decoration while the bottom bracket has 48 pieces applied. The metal, silvered, etched dial with gold wash was probably purchased from Dold and Hettich in Furtwangen. The metal hands have gargoyle type figures etched in them. The theme from the oil painting (which is painted on wood) is taken from a poem by Viktor von Scheffel (1854) and referred to as "The trumpeter from Säckingen." Säckingen is a city in the Black Forest. The case for this clock was designed by R. Bichweiler of Furtwangen and made by Augustin Tritschler of Furtwangen, the most famous cabinetmaker of Black Forest clocks at this time. This is certainly one of the most elaborately made clocks from the Black Forest region. The same model clock sits in the Historical Clock Museum, Furtwangen, Black Forest, for your viewing pleasure. See other views of this clock on color page 48.

See page 8 for poem.

The metal, etched, silvered dial. Gold wash is in the etched part of the dial. Note the hands with gargoyle figures. The gong train of the movement is wound in the center. The trumpeter portion is on the left side with the gong strike wound on the right side looking at the front of the movement.

Two views of the music box located in the base of the Sheffel-Uhr trumpeter clock.

Four-horn trumpeter shelf clock with original bracket made by E. Wehrle about 1890. Etched, silvered dial. Plays one tune. Case is walnut and measures 33 inches tall. The bracket is 14 inches tall.

Smaller shelf trumpeter clock made by E. Wehrle with a nine-horn, two-tune musical movement. The case measures 36'' x 16'' and is made of walnut. Small fluted finials at the top.

Trumpeter Clocks 225

The clocks that follow are the two or three note trumpeter style that are not labeled or stamped with any factory name, but are collectible Black Forest clocks.

A three-note, five foot long, trumpeter clock with pheasant and deer and tree motif. Enamel numerals. Bone hands. The carving was extremely well done.

A nice architectural style shelf trumpeter clock with brass movement that runs three days. When the clock strikes, the trumpeter sounds off on two notes from brass pipes. Circa 1890. Oak case.

226   Black Forest Clocks

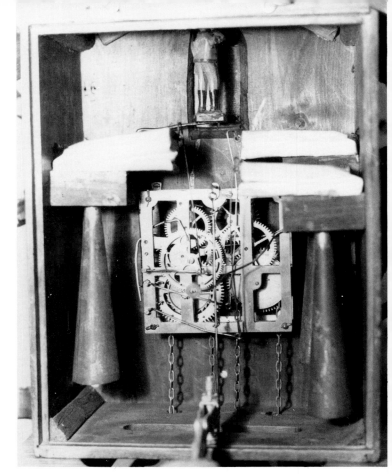

The cast brass movement with three metal pipes and cuckoo type bellow. The pinned plate movement is earlier than the type held together with a nut.

Wall hanging, three note trumpeter clock. Thirty-hour, cast brass movement which is not stamped with any maker's initials. Case has the hunters motif with the deer head or stag, horn with rifles and pouch. Bone hands. Case carving is made from linden wood. The clock sounds three short notes at the half hour and also at the hour. At the hour, 6 o'clock for example, the clock will sound eighteen notes.

The carved and painted wood trumpeter figure, which is original.

Trumpeter Clocks 227

A plaster-type painted trumpeter figure from a two-note unmarked model.

A two-note trumpeter movement with bellow pipes.

A gong used on a two-note shelf trumpeter.

228 Black Forest Clocks

A wall hanging two-note trumpeter clock with an unmarked movement. Bone hands. One-day movement. Circa 1890.

Label in Spanish, translates: "Trumpeter clock Number 2 Germany"

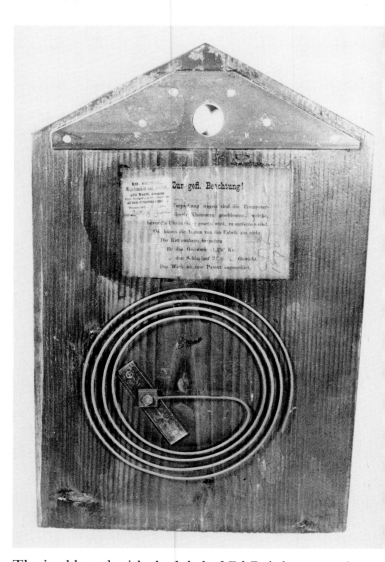

The backboard with the label of Ed Reinke, a watchmaker and jeweler in Chicago. Also an instruction sheet written in German for setting up the clock, removing clamps, etc. The movement is patented.

CHAPTER 10

Small Clocks and Their Makers

JAKOB HERBSTREITH, Hinterzarten

Jakob Herbstreith, nicknamed "Jockele," was a clockmaker in Ravenna Hinterzarten-Breitnau. He was the first to make the small porcelain wag clocks today referred to as the Jockeleuhr or "Jockele." The story of the name's origin is interesting. Jacob's father was also named Jacob, so they gave the son (who made the clock) the nickname of Jockele to avoid any confusion when Jacob was being called through the house. That is how the clock got the name of Jockele. The wood plate, brass gear movements in these clocks were patterned after the style of the larger shield clocks. However, they were much smaller in size, 3⅛ inches high, 2⅜ inches wide, and, depending on the strike, 2 to 2⅜ inches deep.

Time-only, time and alarm, time and strike, and time, strike and alarm movements were made. There were also Jockele movements made that were 3½ by 3½ inches. In these the time train and strike train of gears were side by side. They are also known to have been made with cuckoo bellows. No cuckoo bird was present because the clocks were too small. The front of the clock is porcelain and mounted on a wood plate. The porcelain fronts, mostly with separate porcelain dials, were painted with designs of flowers or small landscapes.

There were different shapes of the Jockele, including the teardrop style, the baroque shape, a picture frame style and others. Most of the small Jockele clocks will fit easily into the palm of one's hand; the first one was made around 1790. The Jockele movements would run for 24 hours. The fronts of the clocks could have been made with the lacquered wood fronts, brass front with porcelain dial, porcelain front with porcelain dial, and they were also carved. The dials were made in these sizes: 6.5 cm. (about 2½ inches), 7.6 cm. (about 3 inches), 9.4 cm. (about 3¾ inches) and larger.[1]

A small Jockele clock with alarm. The porcelain front measures five inches in length. With its porcelain dial, the front is richly painted in red, blue, green, and other colors. The movement is wood plate with brass gears and steel pinions and will fit in the palm of your hand. Circa 1860.

230 Black Forest Clocks

The small time-only Jockele movement, wood plate, brass gear, steel arbor which measures about 3 inches tall.

Side view of the Jockele movement.

Porcelain Jockele on the wall. Time-only movement with brass gears and steel arbors and wood plates. Porcelain front is a little less than 4 inches long. Circa 1850.

This clock shows a more delicate side of the Black Forest Clockmaker. This little porcelain clock, only 5½ inches high, is referred to as the "Jockele," and was made by Jakob Herbstreith of Hinterzarten, Black Forest. This clock is referred to as the teardrop style. There was also the baroque shape and they were made with picture frames, etc.

Jakob Herbstreith, oil painting on zinc, artist unknown. (Historical Clock Museum, Furtwangen, Antique Clocks Publishing Archive).

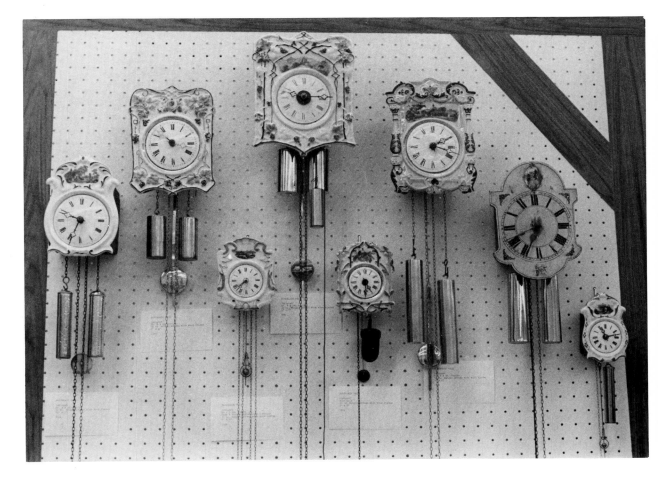

A group of wag on the wall clocks. The three smaller clocks are called the Jockele style. (Photographed at the NAWCC national convention in Cleveland, Ohio, 1986.)

A small porcelain front wag on the wall in a Baroque shape, time-only wood plate movement. Clock measures 4 inches in length. Circa 1860.

JOSEF SORG, Neustadt

Josef Sorg was born May 5, 1807 and learned the clock trade from his father, an innkeeper and part-time clockmaker. Josef made most of his clocks in the mid-19th century. Most of these clocks were time only, however they were also made with strike and alarms. The smaller wood plate, brass gear movements are no longer than about 2½ inches high, unless they are also strike and alarm. The early movements were weight driven with a rope, not a chain. The brass fronts were usually no larger than 3 to 3½ inches in length. The Jockele porcelain fronts can be 3 inches in length to about 5 inches in length. Other clockmakers also copied and made this type of small "Sorguhr." They named them Sorguhr after the inventor. Some are signed "J.S." Josef Sorg also found time to be the Mayor of Neustadt from 1850-1852.

Other clockmakers who were known to make these small types of clocks were Josef Koch of Lenzkirch, Josef Pfaff of Triberg, and Dominik Spiegelhalder of Neustadt. These people were considered to be small clockmakers, not in the business of making larger clocks. These clockmakers had more help in their shops and could make ten to eleven of this type of clock per week.[2]

Footnotes

[1] Berthold Schaff, *Schwarzwalduhren*, pp. 116-123, Verlag Karl Schillinger, Freiburg, 1983.

[2] Berthold Schaaf, *Schwarzwalduhren*, pp. 123-126, Verlag Karl Schillinger, Freiburg, 1983.

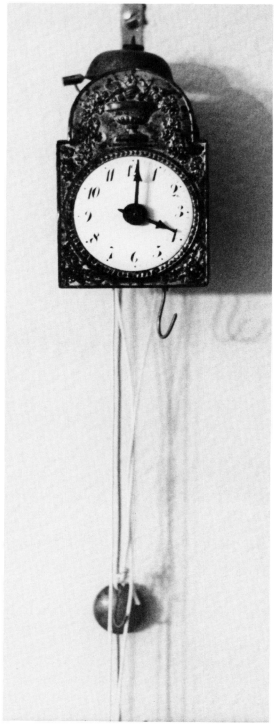

A small brass wag or "Sorguhr" with brass front and enamel dial made after 1800 by Josef Sorg. Time-and-strike wood plate movement with brass gears and wood arbors. The clock measures 9 centimeters (about 3 inches) in length.

A very small clock, only 3½ inches tall, patterned after clocks made by Josef Sorg of Neustadt. It will definitely fit in the palm of one's hand.

Small brass wag on the wall clock referred to as a Sorguhr. Brass front is 9 centimeters (about 3 inches) long. Time-and-strike weight driven movement. Wood plates with brass gears and wood arbors. Circa 1840. Hands are not original.

CHAPTER 11

Organ and Flute Clocks and Their Makers

IGNATZ BRUDER, Waldkirch

Ignatz Bruder was born in Unterharmersbach bei Zell on January 31, 1780. He grew up in poor and disadvantaged surroundings. Ignatz wanted to learn hand work in the arts and craft business, but his parents had very limited funds and could not afford this education. So he worked as a construction worker in Nancy and Mirecourt, France, around 1797. In these cities small organ grinders were built. It appears that Ignatz learned the trade of building organs in the French workshops. It is not known exactly how long he stayed in France, though it was probably about two years. In the church register in Simonswald it is written that he went to those two French towns and studied and learned the organ making trade. Several priests from St. Peter helped him with his learning expenses. He also worked at construction in Simonswald and repaired clocks in the winter.

Ignatz also had close connections with an organ builder named Mathias Martin (1759-1825) who learned the trade at the famous Silbermann School and settled in Waldkirch in 1800. Martin built the organ in the church of Zell when Ignatz was only 13 years old. It seems Ignatz Bruder may have helped him with this task.

In 1806 Ignatz Bruder sold his first successful flute clock. In 1816 he had acquired a two story house with two workshops and sleeping quarters for his workers. He expanded even further in 1823, eventually becoming a very wealthy, respected man. His sons, Andreas and Xavier, completed their apprenticeship as music box makers in 1829 and became a valuable asset to the flute clock and organ grinder firm.

In 1829 Ignatz completed a 284 page handbook, with 24 drawings on organ building; it also included mechanical player organs. He wished to pass this skill on to his children and grandchildren. Ignatz settled in the city of Waldkirch in 1834.

The movement of a Black Forest flute clock with the call of a blackbird, made about 1850. (Gerd Bender, *Die Uhrmacher des hohen Schwarzwaldes und ihre Werke, Band 1*, Verlag Müller, Private collection, H. Dold, Furtwangen).

Flute clock movement by Jakob Streifer. Wood plate, eight-day running. (B.Schaaf, Schwarzwalduhren).

Side door of the flute clock by Jakob Streifer. List of eight songs that are played and the signature of Jakob Streifer in Furtwangen. (B. Schaaf, Schwarzwalduhren).

Here he made flute or organ clocks and small organ grinders for market use. In the last years of his life he made church organs, though he was not very successful with these. Ignatz died in 1845.

His sons took over the business and in 1864 built a large factory building in Waldkirch. The firm Gebrüder Bruder was started and in the process made this city famous. Waldkirch became the city of organ builders. The firm also began to make larger show organs for carnival use. The grandsons of Ignatz Bruder eventually started two separate companies called Wilhelm Bruder Söhne and Ignatz Bruder Söhne, the latter devoted strictly to making grinder organs. In 1880 they went out of business, as competition became stronger.

RUTH & SOHNE, Waldkirch

Another factory located in Waldkirch was Ruth & Söhne. Born in 1813, Andreas Ruth was an apprentice and learned about organ clock manufacturing in Furtwangen and later from Xavier Bruder in Waldkirch. He started his own company in 1841, and by 1879 his firm was just as famous as the Bruder Company.

It can be said that music boxes were made by the clockmakers who were always experimenting. Later on, a profession developed out of this and there were clockmakers that called themselves organ clockmakers. Flute clocks or organ clocks (those without any form of animation), were basically built between 1770 and 1820. After 1820 organ clocks were made more elaborately with full technical, orchestrated, and animated features. It is hard to date flute clocks exactly, because many are unsigned by any maker. Organ clockmakers made these clocks in variations that went on and on. Some had a small number of pipes, perhaps thirteen or fourteen with six or eight tunes. Others had fifty or seventy-five pipes with twelve tunes. Some came with more than one pin barrel and were interchangeable. Some had animation with three or four moving figures, and some had as many as or more than twenty. There were no limitations, it seems, in the organ clock makers imagination.[1]

Footnotes
[1] Gerd Bender, *Die Uhrmacher des hohen Schwarzwaldes und ihre Werke*, pp. 454-474. Verlag Müller, Villingen, 1979 revised.

Organ and Flute Clocks and Their Makers 235

Flute clock in a floorstanding or hall case made by F.X. Scherzinger in Furtwangen about 1875. It has twenty wooden pipes for the music and plays twelve tunes. The movement and flute movement were made by Karl Blessing, Unterkirnach, about 1825. The case for the clock was made at a later date. (Deutsches Uhrenmuseum, Furtwangen/Schwarzwald. Photo, Callwey Verlag, München, Germany).

Floorstanding flute clock with moving figures made by Ignatz Bruder in Simonswald. The clock has 83 wooden pipes for the music with eight musical tunes. The movement with hour chime runs for 24 hours. The music roll guides the figures and their movement. (Gerd Bender, *Die Uhrmacher des hohen Schwarzwaldes und ihre Werke, Band 1*, Verlag Müller. Private collection, E.I. Amrein, Basel/Schweiz.)

CHAPTER 12

Animation Clocks

This section shows a wide variety of Black Forest animation clocks, but all of the clocks are not labeled or stamped by any clock factory name. These clocks are highly sought after by collectors. These types of clocks do more than just tell time. Some have carved figures of men placed on top of the clock case that move when the clock strikes. Others have moving eyes within the dial and some have soldiers that move back and forth across the clock case as if to guard the castle-style case. Many of the figures, like the potato eater, move their arms, mouths, and legs as you will see in this chapter.

The Black Forest magician clock. The case measures 25 inches with a figure at the top of the case, that figure being a magician. This figure works hard: about every fifteen minutes he raises both arms to show the dice or deck of cards to his audience, fifteen minutes later he raises his arms again to show the dice or deck of cards have disappeared and have been replaced by a pea. Circa 1875.

Animation Clocks 237

The carved wood figure holding his bowls for the next act.

The brass movement. The lifting wires connect through the figure and table legs.

The backside of the painted standing figure.

238 Black Forest Clocks

A Rat Eater animation clock, made about 1870. The case is inlaid with a metal painted dial. The figures' eyes move with the swing of the pendulum. The figures' arms moves to his mouth, which opens as if to eat the rats on the plate. (Cleveland, 1986 NAWCC National Convention)

A Potato Eater, or "Knodelfresser," clock which measures 22 inches tall, with the figure at the top representing a person from the French Revolution. Made about 1870. The clock strikes at the hour. The figure lifts its arm to its mouth and the mouth opens as if a potato was being eaten. The eyes of the figure also move with the swing of the pendulum. Phillip Haas and Son, in St. Georgen, were well known as special makers of the Potato Eater around 1880.

Animation Clocks 239

The figure painted and carved from wood. The right arm and eyes move.

The unrestored brass movement of the Potato Eater.

A wood shield type clock with a wood plate movement with brass gears, steel arbors, and a painted dial. The figure at the top is that of a tailor, hard at work ironing. It has been stated in the publication, *The Watchmaker and Jeweler*, June 1875 issue, "The most curious branch of art is, without exception, the manufacturers of automation clocks and the ingenuity of the Germans has produced some wonderful mechanical clocks."

240   Black Forest Clocks

A double blinking eye picture frame clock. The eyes of both children move with the swing of the pendulum. The painting and dial of this clock is the reverse glass type, although many of the scenes were painted on metal in oil and had enamel dials. These types of blinking eye clocks were made from the middle to the end of the 19th century. Any object that had eyes was used for the blinking eye motif—people *and* animals.

The back of the wood plate movement used for the blinking eye clock. The wire from the verge extends to the wire apparatus above the movement which is connected to the enamel eyes that move with the swing of the pendulum.

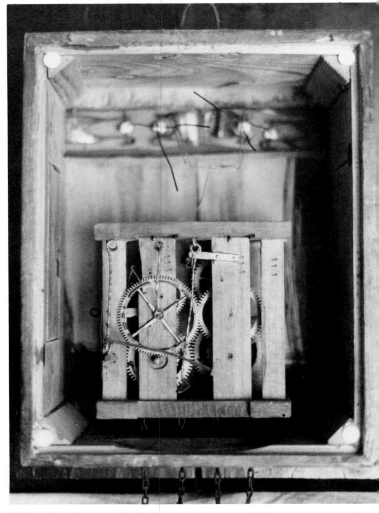

The simple wood plate movement with brass gears and steel arbors. A wire leads from the escapement to the wire apparatus above the movement which is connected to the wires that hold and move the eyes. This movement runs one day.

Animation Clocks

Black Forest clock, made about 1880. Spring driven movement runs about two days. Roots are applied to this case. The monk bell ringer appears at the hour and rings 36 times. Photographed at a NAWCC convention, San Jose, CA.

The bell ringer, who appears at the hour. A monk or priest. Note the tree roots at the bottom.

Monk ringing bell alarm clock. Three weight, thirty-hour, wood plate movement, with time, strike, and alarm. Case is 23 inches tall by 12 inches wide. Circa 1865. Enamel dial with original hands. (Cleveland, 1986 NAWCC National Convention)

Black Forest animated dumpling eater. The arm and mouth move. Eight-day brass movement. Case is 23 inches tall, with inlay and metal dial. The paint on the carved figure is original. Metal hands are original.

A small carved desk clock which measures 10 inches tall. The teeter-totter moves back and forth over a barrel. Thirty-hour brass movement similar to a Lenzkirch style. (Cleveland, 1986 NAWCC National Convention)

An example of a wood plate time-and-strike movement with brass wire attached to the verge to move the eyes in a blinking eye clock. Circa 1860. The gears are brass with steel arbors.

Animation Clocks

An automated figure clock. The King, at the hour, moves his left arm, holding a bottle of refreshment, and pours himself a drink in a glass in his right hand. The right arm holding the glass is raised to his mouth which opens and accepts the refreshment. The firm of Crisp Brothers on Brick Lane in Shirlfield is painted on the metal dial.

The movement with brass wires that activate the arms and mouth of the King connected through the hollow figure.

The label of Crisp Brothers in London is pasted on the backboard of the clock. They handled many different types of clocks. June 11th, 1878 can also be seen written on the backboard.

Soldier and cuckoo shelf clock with columns on the front of the case which resembles a small fortress or castle. Circa 1900.

The soldier and cuckoo clock movement.

Mantel animation clock with the figure of a scissor sharpener on the top in an oak case, made about 1860. Spring driven movement. (Deutsches Uhrenmuseum, Furtwangen/Schwarzwald, photo Callwey, Munchen, Germany).

Animation Clocks 245

Blinking eye picture frame clock. The eyes move with the swing of the pendulum. Circa 1870.

The brass, unmarked, spring driven movement, with the eye apparatus above.

A shelf cuckoo with soldier at the bottom of the case that moves across, turns around, and moves back the other direction with the swing of the pendulum. Circa 1900. The case measures about 20 inches tall.

CHAPTER 13

Black Forest Regulators

The regulators depicted in this section are patterned after the Vienna Regulator style of clocks, however these are German made and are not stamped or labeled with any factory firm. They were made by Black Forest clockmakers, mainly in factories, and were mass produced from 1870 to 1900.

Miniature spring driven RA style. Unsigned. 8-day movement. False grain finish with black trim on case that measures 17 inches in length. Circa 1875.

Spring driven miniature. Unsigned. RA pendulum insignia. Unusual case. 8-day movement is not stamped with any factory. The door is not made of wood, but from a pressed or molded type of material which is very sturdy. Enamel dial. Circa 1875.

Black Forest Regulators 247

Miniature regulator 43½ cm in length. Case is bronze. Metal pendulum rod with engraved bob. Spring driven 3-day movement is not stamped with a factory mark. Circa 1870.

Carved, spring driven miniature. Unstamped movement. Length of case is 14". Dial is 3" in diameter. Pendulum is RA type. No chime. 8-day movement. Circa 1890.

Miniature RA, spring driven without strike works, no stamp on movement. Case is 35 cm long. Two piece dial is 6 cm in diameter. The case is black with flowered pattern and missing 2 bottom finials. Eight-day movement. Circa 1890. (Antique Clock Gallery, Long Beach, CA)

RA style clock made about 1870. Walnut case with black trim on the door, top, and bottom of the case. Unsigned 14-day movement.

Freeswinger style made with a running duration of three weeks. Walnut stain. Pendulum is enclosed by a glass door. Also referred to as a "Berliner." Unstamped movement runs 14 days. Circa 1885.

Spring driven regulator. Length of case is 36 inches with full columns. Etched dial and pendulum bob. Glass only in the front door. Translated inscription says, "So goes time forever." Walnut finish. 17-day movement, not stamped with any factory mark. Gong chime with half-hour strike. Circa 1880-1890.

Factory-made spring driven clock with a music box for the strike portion of the 8-day movement, which is not marked with a factory stamp.

Miniature regulator, probably made about 1875. Inlay on the case. Unsigned. This is one of the nicest and most intricately detailed cases made. 8-day movement.

Spring driven regulator. Walnut case with brass decoration (30 pieces) is 104 cm long. 14-day movement. Circa 1890. The movement is not stamped with any factory mark.

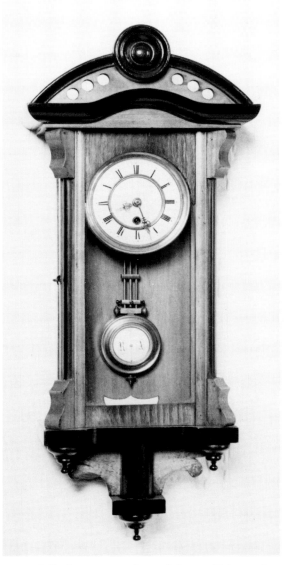

Miniature spring driven RA style. Case is 40 cm in length. Dial is 7 cm in diameter. 5-day movement. Circa 1880. Movement is not stamped to indicate which factory made this clock.

Carved wall clock with RA pendulum, with a grouse at the top. Spring driven, time only. Case is 19 inches tall by 10 inches wide. Circa 1880. 7-day movement. Has the Black Forest "American styled movement" similar to movements made by C. Jerome & Co. in the U.S.

Smaller spring driven regulator without strike works. Figure 8 style with carved front door. 14-day movement is not stamped with a factory symbol. Circa 1870.

APPENDIX 1

Reproduction of pages from the 1904 Twelfth Annual Catalog, St. Louis Clock and Silverware Company

The following four pages are reproduced from the 1904 Twelfth Annual Catalog, St. Louis Clock and Silverware Company. Permission to reprint was granted by Robert Spence, former President of American Reprints Book Company, now retired. Most clocks pictured on these pages are taken from the Gebruder Kuner factory.

BLACK FOREST CLOCKS. (Schwarzwald) GERMANY.
IMPORTED DIRECT FROM THE BLACK FOREST (Schwarzwald), GERMANY.

The carving of **these Clocks** is done by hand by the natives of the Black Forest and is especially fine. The figures are accurate and life-like. The **fronts of these Clocks** are made of **three pieces** of wood of opposite grain and are set together to prevent warping and cracking. The movements are made of the best tempered steel and brass with screw pillars, patent ratchet wheels, new escape wheel, and are well finished and adjusted. The same are **guaranteed to be correct timekeepers.** Every **Clock** that leaves the factory has been run and thoroughly tested, and is warranted, if properly used, to be a good timekeeper. Should a case crack or split, it will be exchanged or repaired without charge, if transportation charges are prepaid. **All genuine** Black Forest Cuckoo Clocks are stamped with this Trade Mark.

TRADE G. K. MARK.

No. 44. CUCKOO CLOCK, $7 50
Height, 19 inches; width, 12 inches. Half-hour strike and call.
German **Walnut** or **Oak,** with Inlaid Ash, Ebony and Mahogany ornaments.
The front in the case of this Clock is made of a single piece of wood and the movement is of a smaller size.

No. 561. CUCKOO CLOCK, $9 00
Height, 18 inches; width, 12 inches; dial, 4 inches. Half-hour strike and call.
Imitation **Walnut** or **Oak.**
The front of the case of this Clock is made of a single piece of wood, and the movement is of a smaller size.

No. 55. CUCKOO CLOCK, $9 30
Height, 21 inches; width, 14 inches; dial, 5½ inches. Half-hour strike and call.
Solid **Walnut** or **Oak** case, with inlaid Ash, Ebony and Mahogany ornaments.
Large size **G. K.** movement. Front of case made out of three layers of wood. **Guaranteed.**

No. 37. CUCKOO CLOCK, $11 00
Height, 23½ inches; width, 15½ inches; dial, 6¼ inches. Half-hour strike and call.
Solid **Walnut** or **Oak,** with Cherry Ornaments.
Large size **G. K.** movement. Front of case made out of three layers of wood. **Guaranteed.**

No. 28. CUCKOO CLOCK
(Log Cabin), $11 70
Height, 20 inches; width, 14 inches; dial, 5 inches. Half-hour strike and call. **Walnut only.**
Deer's head and horns are carved out of wood also.
Large size **G. K.** movement. **Guaranteed.**

No. 58. CUCKOO CLOCK, $10 70
Height, 21 inches; width, 14 inches; dial, 5½ inches. Half-hour strike and call.
Solid **Walnut** or **Oak,** with inlaid Ash, Ebony and Mahogany ornaments.
Large size **G. K.** movement. Front of case made out of three layers of wood. **Guaranteed.**

Cuckoo Clocks From the Black Forest of Germany

No. 33. Cuckoo Clock, $13 90

ght 19 inches, width 14 inches, dial 5¼ inches, alf-hour strike and call. Solid **Walnut** or **Oak**. arge size **G. K.** movement. Front of case made t of three layers of wood. **Guaranteed.**

No. 6. Cuckoo Clock, $17 30

Height 23 inches, width 14 inches, dial 6 inches, half-hour strike and call. Solid **Walnut** or **Oak**. Large size **G. K.** movement. Front of case made out of three layers of wood. **Guaranteed.**

No. 4. Cuckoo Clock, $15 50

Height 19 inches, width 14 inches, dial 5¼ inches, half-hour strike and call. Solid **Walnut** or **Oak**. Large size **G. K.** movement. Front of case made out of three layers of wood. **Guaranteed.**

No. 47. Cuckoo Clock, $16 00

ght 20 inches, width 13 inches, dial 5¼ inches, lf-hour strike and call. Solid **Walnut** or **Oak**. rge size **G. K.** movement. Front of case made t of three layers of wood. **Guaranteed.**

No. 24. Cuckoo Clock, $16 50

Height 21 inches, width 15 inches, dial 5¾ inches, half-hour strike and call. Solid **Walnut** or **Oak**. Large size **G. K.** movement. Front of case made out of three layers of wood. **Guaranteed.**

No. 67. Cuckoo Clock, $18 00

Height 23 inches, width 14 inches, dial 6 inches, half-hour strike and call. Solid **Walnut** or **Oak**. Large size **G. K.** movement. Front of case made out of three layers of wood. **Guaranteed.**

QUAIL and CUCKOO and CUCKOO ONLY. FROM BLACK FOREST, GERMANY

No. 46. Quail and Cuckoo Clock, $27 30

Height, 21½ inches; width, 16½ inches; dial 6⅝ inches. Quarter-hour strike and call.
Solid **Walnut** or **Oak**.
Large size **G. K.** movement. Front of case made out of three layers of wood. **Guaranteed.**

No. 7. Quail and Cuckoo Clock, $22 30

Height, 25 inches; width, 17 inches; dial, 6⅝ inches. Quarter-hour strike and call.
Solid **Walnut** or **Oak**, with inlaid Ash, Ebony or Mahogany ornaments.
Large size **G. K.** movement. Front of case made out of three layers of wood. **Guaranteed.**

No. 8. Quail and Cuckoo Clock, $25 00

Height, 21 inches; width, 16 inches; dial, 6⅝ inches. Quarter-hour strike and call.
Solid **Walnut** or **Oak**.
Large size **G. K.** movement. Front of case made out of three layers of wood. **Guaranteed.**

No. 81. Cuckoo Clock, $21 00

Height, 22 inches; width, 16 inches; dial 5¾ inches. Half-hour strike and call.
Solid **Walnut** or **Oak**.
Large size **G. K.** movement. Front of case made out of three layers of wood. **Guaranteed.**

No. 83. Quail and Cuckoo Clock, $46 00

Height, 28 inches; width, 16 inches; dial, 6¾ inches. Quarter-hour strike and call.
Solid **Walnut** or **Oak**.
Large size **G. K.** movement. Front of case made out of three layers of wood. **Guaranteed.**

No. 56. Cuckoo Clock, $22 00

Height, 23 inches; width, 15 inches; dial, 5¾ inches. Half-hour strike and call.
Solid **Walnut** or **Oak**.
Rabbit, bird and deer's head decorated in natural colors. Imitation glass eyes in rabbit's and deer's head.
Large size **G. K.** movement. Front of case made out of three layers of wood. **Guaranteed.**

Our Direct Importation Novelty Clocks.
TRUMPETER, STANDING CUCKOO, MUSIC AND MINIATURE CLOCKS.

No. 10. Cottage Cuckoo, $6 00
Height, 13 inches; width, 6¼ inches; depth, 4¼ inches. One-day spring movement. Half-hour and hour cuckoo-call; 3-inch dial.
Solid **Oak**, brass trimmed.
Imported novelty.

No. 17. Trumpeter Clock, $18 00
Height, 25 inches; width, 17 inches; dial 6⅝ inches. Two weights. Half-hour Bugle-call of the Trumpeter. One door for Trumpeter.
Solid **Walnut only,** with inlaid Ash, Ebony or Mahogany ornaments.
Front of case made out of three layers of wood.
Guaranteed.

No. 55½. Joker Music Alarm Clock, $7 00
Gold gilt front and handle, nickel-plated frame and glass sides. Height, 7 inches; width, 5 inches.
This clock plays two tunes for about ten minutes, instead of ringing of a bell.

No. 7434. Kitchen Clock, $3 70
One-day time (spring). White enameled Delf painted case. Height, 13 inches.

No. 61. Miniature Clock, $1 30
Height, 7 inches; width, 4½ inches; front and dial made out of burnt wood.

No. 50. Cuckoo Clock with Music, $20 00
Height, 19½ inches; width, 12½ inches; dial 6 inches. Half-hour strike and call, with musical attachment, playing a tune at the same time.
Walnut only.
Large size **G. K.** movement. Front of case made out of one piece of very heavy solid Walnut.
Guaranteed.

No. 62. Miniature Clock, $1 20
First quality.
Height, 7 inches; width, 4½ inches.
Fitted with new patent celluloid dial.
In **Walnut** or **Oak** colors.

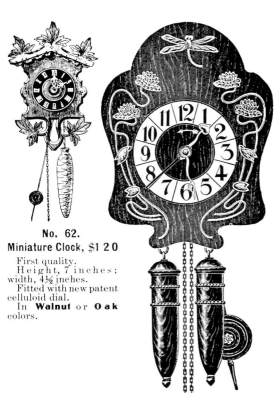

No. 334. German Alarm Clock, $8 50
Height, 11 inches; width, 7 inches. The loudest Alarm Clock made.
Porcelain Dial. Aluminum trimmings. Green finished **Oak** case and weights.
Guaranteed.

APPENDIX 2

Carved Clocks with French Style Movements

The clocks shown in this section were possibly made in the Black Forest area, however the movements and cases are not stamped with any clock factory. Most have the round French style movements. This writer has seen other examples of this style of clock that were stamped with a Black Forest clock factory. Some of these clocks are exquisitely carved by master artisans.

Black Forest carved clock with 14-day French movement made with brass plates and gears and steel arbors. Ram and dog motif exquisitely carved. Dark finish on the case. Circa 1880.

Appendix 2 257

Exquisitely carved clock with hunter and deer at top and quail at the bottom, rabbit and squirrel on sides. 14-day French movement. Dial has porcelain numbers. Circa 1880.

258 Appendix 2

A small, carved mantel clock with a French time only, 14-day brass movement and porcelain dial. The case is 11 inches tall. The carving is from one piece of wood and was made somewhere in the Black Forest region, but there is no maker indicated.

Carved clock with French style movement that runs for 14 days. Brass hands with porcelain numerals. Eagle and deer motif. Circa 1890.

APPENDIX 3

The Black Forest Clock Industry 1875-1892: From a Horological Magazine's Viewpoint

The following is a view of the Black Forest clock industry taken from various sections of periodicals: *The Jewelers Circular and Horological Review*, from the years 1884, 1886, 1887, 1888, 1891 and 1892; and *The Watchmaker and Jeweler*, 1875. The information is included in this book because it gives a contemporary look at the Black Forest clock industry about 100 years ago from the periodical standpoint. It was provided by *Horological Data Bank, National Association of Watch and Clock Collectors Museum, Inc.*

From *The Watchmaker and Jeweler, June 1875*

CLOCK-MAKING IN THE BLACK FOREST

A French contemporary gives the following interesting account of clock-making in the Black Forest, which industry is, at the present day, concentrated at Dittishausen Eisenbach, Furtwangen, St. Georgian, Lenzkirch, Neustadt, Freiberg, Villingen, and Rohzenbach. Essentially a domestic occupation at its origin, and employing whole families, it is only thirty years since watch and clock-making have been carried on in factories. Lenzkirch produces a great number of what are called Parisian watches, the zinc for the cases being sent from Paris and gilded here. The cases, wheels, and all the necessary items are made separately, and the work is divided into various kinds; the separate pieces are carried into the workshops, where they are put together, when the clocks are successively mounted, tested, and regulated. The different sorts of clocks manufactured in the Black Forest comprise—the clocks with weights; clocks in cases, amongst which are 12-hour clocks, 24-hour clocks, 8-day clocks, clocks which go for a month; tower clocks; regulators; the spring clocks of English and French construction; ship clocks; and figure clocks, amongst which may be classed cuckoo and trumpet clocks.

The importance of this branch of industry will be seen by the following figures: In 1871 the Black Forest numbered 1,429 free manufacturers of clocks and watches, employing 7,526 hands, independent of women and children, who were occupied in the small details; 13,000 persons lived by this industry alone. The number of articles produced had risen in 1874 to 1,800,000, of which 100,000 were of first-class workmanship. The total represented a value of ten millions of florins.

The most curious branch of this art is, without exception, the manufacture of automaton clocks, and the ingenuity of the Germans has produced some wonderful mechanical clocks. The great musical clocks are called orchestrals. The development of this art is due to Vaucanson, who made the first at Blessing. In the manufacture of these clocks the work is not divided, at least for the essential parts. The masters execute nearly all in their workshops, with the exception of the metallic pipes. Last year there were 32 masters and 224 hands employed in making these clocks. The great factories of orchestral clocks produce instruments of five and six registers, of which the price varies from 1,000 fl, to 20,000 fl. These clocks are in great demand in America and Russia.

From *The Jeweler's Circular and Horological Review*, October 1884

CLOCKMAKING IN THE BLACK FOREST

The royal commission on Technical Education has published a report on their investigations, from which we make the following extract in reference to clockmaking in the Black Forest.

By means of lotteries and further grants, the Gewerbe-Halle was erected and opened in 1874, and the school was transferred to the new building, and was re-constituted and formally opened in June, 1877. The government grant for this purpose was 650 pounds. The school is managed by a local council chosen from the surrounding districts, and consisting of eight persons. The annual budget is 360 pounds, of which sum the Province of Freiburg contributes 25 pounds and Villingen 50 pounds. The school is held in two small rooms, very ill-suited for the purpose, and a new building is in course of erection by the local Gewerbe-Verein (trade society) at a cost of from 1,250 pounds to 1,500 pounds. There are at present eighteen pupils. Most of them are admitted free, and many are supported by exhibitions which are given by the neighboring towns, and may amount to 20 pounds per annum. The school fee is 1 pound per annum.

The students must have been previously engaged for at least two years in practical clockmaking before they can be admitted. The course lasts one year. In the morning they have theoretical studies, consisting of geometry, arithmetic, algebra, physics, mechanics, technology of watch and clockmaking geometrical drawing, projection, technical drawing, bookkeeping and commercial arithmetic; together with thirty hours' practical work per week in the shops in the afternoon. The lathes and tools seem excellent of their kind, but the accommodation was very cramped.

By far the most important manufacture of the Black Forest is that of small carved clocks, many of them with musical accompaniments, known as cuckoo-clocks. In Triberg and a group of small towns and villages surrounding it, known as the clock country, 13,500 people, the population of upwards of ninety parishes, are engaged in clockmaking. We went to Schonach, a village in the hills above Triberg, where almost every cottage is the home of a clockmaker. The houses are large and substantially built. They are generally a considerable distance apart, and are surrounded with meadow-land and pastures.

In addition to the home workers there are three or four small factories in which the wheels and other parts of clocks (Uhrenbestandtheile) are made. In one of these, visited by us, there were about five or six workmen employed and an apprentice. The power was derived from a small overshot wheel. There was a wire-cutting and straightening machine for preparing the pinions, a lathe for shouldering and tapering the pinions, some fine drills, a tooth-cutting machine, a shaping machine for the wheel teeth and a small press for driving on the bosses, etc. The machinery was ingenious and well adapted for its purpose. The rough castings made in the village were excellent. We were told that the men working piece work in the factories could earn from two to three marks per diem, though a very small proportion of the workmen earn the latter sum.

In a second factory we visited subsequently, the movements made here were being fitted together and made into clocks. Carpenters were making the rough framework to contain the wheels, and others were preparing the varnished outer cases. The clock passed from hand to hand, one man adding the winding barrel, another the train of wheels, a third the escapement, and another placing the works into the case. The clocks were being packed up for sale in another part of the works. They seemed all pretty much of one pattern and of a very common description. The proprietor informed us that he produced clocks of exactly the same kind as those that were made in the cottages.

They went mainly to various parts of Europe. None now go to England or to America. The Americans not only supply themselves with cheap clocks, but they have driven out the Black Forest maker from the English markets. The people employed at clockmaking live, we were told, very poorly—mainly on potatoes—though we saw soup and meat in some of the cottages. The cottages are very large; some of them contain as many as twelve families. Many of the people have only one or two rooms and a bit of land, probably about twenty perches, for which they pay from ten to fourteen marks rent per month. An old gentleman mowing in a very wet meadow was pointed out to us as the Bürgermeister of Schonach. We went into several of the cottages and saw the people at the work. The women were polishing the clock cases, and the men were making the wheel work. We were told here that an industrious worker at home could earn as much as a good workman in a factory. The water power is everywhere most ingeniously utilized, and the water itself is, as in many parts of England, employed for irrigation on the steep hillsides.

BLACK FOREST ART WORK FOR CHINA

December, 1885 (Foreign Gossip column)

Lamy Sons, of Furtwangen, have lately placed on exhibition a complicated specimen of Black Forest handiwork, in the shape of a clock, ordered by a Chinese mandarin. Two birds are enclosed in a tower at the foot of the dial, one of which sings at the half-hour, the other at the full. Two other birds roost in a tower above the dial, one of which sings at one-quarter, the other at three-quarters of an hour. The full-hour bird warbles a Chinese air. The train contains a quarter-hour work, which effects the unlocking of the birds. The mechanism works to a charm, and the whole clock is painted in the Chinese colors. [Carl Konstantin Lamy was a teacher at the Clock and Watchmaker School in Furtwangen (1850). He started his own business in 1864. His family had moved from France to Basil, Switzerland]

November, 1886

The following report of Frank W. Ballou, United States Consul at Kehl, Germany, has just been published by the State Department:

There is scarcely a city in Germany or in Europe, perhaps outside the limits of this continent, where one cannot find clocks from the Black Forest. It is rarely the case that a branch of industry becomes so extended as the above mentioned, and the Black Forest clocks have become typical for the description of the idea conveying the comfort and snugness of a South German home.

Owing to the barrenness of the soil and the custom in the Black Forest region that farms belonging to a father's estate can only revert to the eldest son, the other children of the family, who were and are still, as a general thing, numerous, were obliged either to emigrate or to engage in mechanical labor of some kind, which was made easy and profitable owing to the magnificent forests furnishing the necessary wood.

As is shown by reliable data the manufacturing of wooden ware, such as turners' and coopers' articles, was carried on largely in the time of the Emperor Rodolphus of Hapsburg. Afterwards the making of tin spoons, brushes and other wares began. Towards the end of the seventeenth century farmers who could not maintain their families from the products of their farms began to engage in clockmaking. Poor cottagers without property, but with inventive genius, and dependent upon the labor of their hands, became the patriarchs of clockmaking. Their sons and their apprentices introduced this industry in the districts where it is to-day a flourishing business. The case and movement of these clocks were made with tools of the very plainest description, such as a pair of compasses, a little scroll saw, augers, and a knife. These wooden clocks proved to be a fortunate invention; they found a good sale and awakened the artistic talent of their inventors. The art of clock making was still the monopoly of the patriarchs, their sons and grandsons, but the inventive genius that is inherent to the inhabitants of the Black Forest united the different experiences gathered in various workshops and made use of them, so that the local extent of this clock industry became largely increased.

In the middle of the last century clock making in the Black Forest was already carried on in all the localities where it now exists. The construction of such a wooden clock with heavy weights was very plain. It consisted of three wheels only, a vertical swing wheel put into motion, a balance resembling a yoke, to which were attached several leaden weights in order to regulate the movement of the clock. This construction was, however, abandoned in 1740. It was supplanted by the pendulum clock. At first the pendulum was placed before the dial, afterwards a long pendulum was arranged behind the movement. All these clocks had to be wound up every twenty-four hours. Some of them were made to strike every quarter of an hour; clocks with automatic figures, such as peasants, soldiers that sounded a signal at the end of every hour, etc.; also clocks that recorded the month and days as well as the hours.

About the year 1750 they adopted movements made of wire instead of wood; afterwards metallic wheels were introduced. Since the year 1750 they manufactured very neat little carved clocks with weights, which were called "Jockele-Uhren," from their inventor Jacob Herbstreith. At the beginning of this century clock making had become quite extensive and brought large profits to manufacturers. From 1830 to 1850, however, this industry did not prosper, owing to the stubborn preservation of old shapes and the manufacturing of clocks of inferior quality. With a view of stopping this decline a technical school for clockmakers was founded at Furtwangen, the center of the clock industry, in 1850, under the direction of Mr. Gerwig, who afterwards became one of the builders of the St. Gothard Railroad. Its purpose was to educate persons in the making of clocks, and to have this industry keep up with the times and progress manifested in other countries, and also to introduce the manufacture of watches.

The latter project had to be abandoned owing to foreign competition, but in other respects the influence of the school was soon felt, for since that time they make better and finer clocks, which answer more the requirements of artistic taste. After a twelve years' existence, when the task of the school was considered to be accomplished, it was dissolved; however, professional schools were founded in the principal localities of the Black Forest, in order to give the necessary technical knowledge to apprentices.

Prominent manufacturers also organize exhibitions showing the condition of clock making, and the progress made therein. Although modern, fine clocks are principally made, there are still establishments which produce the traditional old-fashioned clocks, such as they made 150 years ago, with the primitive wheels and the plain wooden dial. This is less due to the adherence of the Black Forest inhabitants to old tradition than to the fact that these clocks are still largely sought for on account of their durability.

At one time the maker made everything that belonged to a wooden clock. But soon the increase of sales rendered a division of work necessary. No sooner did the clockmakers see that their products went off rapidly than they left to others the preparation of those parts which required more care than art. This was the first step towards that astonishing development of this industry, because it gave time and leisure to the artist to entirely apply his talent to new inventions and improvements.

Owing to the increasing demand it became necessary to divide the various branches of the clock industry, but the latter retained nevertheless the character of a house industry, because the workmen are not employed in factories; they are small farmers, having an acre or two of land with some cattle, and their spare time is devoted to making parts of clocks. This system will undoubtedly remain, owing to the peculiarities of the Black Forest and the character of its inhabitants. The division of labor for a Black Forest clock is as follows:

(1) The wood cutter, preparing the wood of beech tress for the case; (2) the case maker; (3) the maker of the plate (shield); (4) the painter; (5) the founder of the bell and wheels; (6) the chain maker; (7) the spring maker; (8) the carver, (9) the dial maker; (10) the decorator of the case, and (11) the maker of the primitive movement.

In the Black Forest there are 92 communities engaged in this industry, with 1,429 independent clockmakers, giving employment to 7,526 operatives. In 1796 these workshops turned out 75,000 clocks; in 1808, 200,000, and in 1880 the total production was 1,800,000 clocks. In the city of Furtwangen were manufactured over 400,000 of these.

The first clockmakers only made a few clocks for the surrounding farms. The favor these clocks met with determined some dealers in glassware and straw hats to take them among their articles. The net profit they realized was so considerable that it excited the envy of the clockmakers, who upon that experience sent the clocks on their own account to the neighboring large cities in Brisgovia and Suabia. They gradually extended their trade, and divided the markets in the different provinces of Germany among each other; some of them even traveled to Asia and Africa. The Black Forest clocks are sent to all points of our globe.

Germany takes all kinds of clocks, from the finest regulators to the plainest wooden clock.

Austria buys only cheap articles, such as chain-work clocks. The high entrance duties are an impediment to the trade.

Switzerland has a predilection for trumpeter and cuckoo clocks for the use of strangers, and drag-spring; clocks for the native population.

England takes trumpeter and cuckoo clocks and regulators. Some years ago there was a good demand for cheap clocks with weights, but this has materially decreased, owing to the sharp competition from American manufacturers.

France bought no clocks for several years immediately after the war. At present they buy many carved clocks, called "Schottuhren."

Belgium and Holland require wooden clocks with bronze frames.

Russia imports a large number of carved regulators, also light day clocks in polished cases, to be used on Russian farms.

Turkey desires mostly cuckoo clocks, with paintings.

Spain and Portugal buy bronze-framed, carved clocks, with weights.

The United States takes trumpeter and cuckoo clocks with painted dials; also many regulators and musical clocks.

The exportation of the clocks to the United States is steady, and will aggregate $50,000 per year. During the summer months the Schwarzwald clock region is visited by many Americans, and nearly every visitor purchases one of these clocks. They are very attractive and appear to be cheap, but in many cases they are made to be sold only, and an attractive exterior may induce many to purchase an almost worthless article.

One cannot be too particular when purchasing one of these clocks, for when the cuckoo will not coo any more, and the trumpeter will not blow another blast, then is their value as curiosities gone, and when, after a few months, they become valueless as timekeepers then are they very poor stock indeed. I have heard so many complaints from people who have purchased these clocks in regard to their general poor quality, that I deem it my duty to make this fact public, and also to inform would-be purchasers that, if they wish to avoid disappointment, they should be very particular where and of whom they purchase, and in no case to purchase of irresponsible parties. A few inquiries will generally disclose the required facts.

September, 1888

To Clean Dirt Wheels, Pivot Holes, etc.

The repairer gets occasionally a very dirty clock to clean, which looks as if its case had been made the dust pit for the storage of the house cleanings. When he has a clock of that kind before him, we would recommend the method of Mr. B. Morgossy, of Neusatz. He says: "I clean the very dirty wheels, pivot holes, chains, etc., of Black Forest clocks for many years in the following manner: I put a certain quantity of benzine in a square or round tin box with joint, lay the wheels, etc., in, and leave them in for about ten or fifteen minutes. They are then taken out, and with a clean, stiff brush, brushed thoroughly; the operator must take care, however, not to come near to an open light, as the benzine will ignite easily.

The wheels, frame, chain, etc., can also be cleaned in another manner. The brush is dipped in benzine and used for brushing the wheels, etc., by which they become very clean. The pivot holes can quickly be cleaned with a cord, which is fastened in the vise, drawn through the hole, and then a few motions up and down will clean the latter. The brass chains are cleaned with vinegar and a spoonful of salt; they are rubbed between the hands until perfectly clean, after which they are laid in clean water, rubbed well, and dried with a clean rag."

October, 1892

Another Wonderful Clock

An Englishman claims to have acquired from a Black Forest maker, for the sum of 16,000 marks, the most wonderful timepiece that has yet been made. This ingenious mechanism, it is averred, contains every thing that is to be found on other clocks and on calendars; and, in addition, it indicates Berlin, St. Petersburg, Madeira, Shanghai, Calcutta, Montreal, San Francisco, Melbourne and Greenwich time.

Every evening at 8 o'clock a handsome young campanologist invites hearers in bewitching tones, to vespers in an electrically illumined chapel, where a devout-looking damsel is dreamily playing the familiar "Maiden's Prayer." On New Year's Eve two trumpeters announce the flight of the old and the advent of the new year. In May the cuckoo makes a welcome appearance; in June the quail comes forth; and on October 1 a magnificent pheasant is ushered in, only, alas, to be ruthlessly shot down by a typical British sportsman, who proceeds to bag his game in the orthodox style.

At day-break a golden sun rises, bells the while merrily tinkling an appropriate German air, "Phoebus Awakes." At full moon the strains of another favorite Teutonic melody, "Sweet and Tranquil Luna," are ravishingly discoursed. Chanticleers, angels, Death with his rattling bones, gray beards, youths and children of both sexes, and other allegorical figures crowd the metaphorical stage of this wondrous clock, which, prior to its purchase by a wealthy Englishman had extorted the almost awe-stricken admiration of the 40,000 residents of the little town of Furtwangen, in Baden, in whose Industrial Hall it had been temporarily exposed to public view.

WANTED BY THE AUTHOR

I have a genuine interest in all Black Forest clocks and I am always looking to buy good examples of the clocks that are shown in this book, especially trumpeter, Beha cuckoos, or nicely carved eight-day shelf cuckoos, Lenzkirch (especially with brass decoration on the case), musical or animation clocks, and Sorguhr or Jockele clocks (all clocks considered). If you have something that you are contemplating selling, I am always interested in making a fair offer.

I am also looking for other examples of Black Forest clocks to photograph for a future revision to this book. If you have something of interest, please contact me for further information:

Rick Ortenburger
P.O. Box 194
Agoura, CA 91376-0194 USA

Bibliography

Reference and Recommended Literature

Arthur, W.J.G. *Music Boxes, A History and Collector's Guide.* Ord-Hume, London.

Bender, Gerd. *Die Uhrmacher des hohen Schwarzwaldes und ihre Werke,* Band I & Band II. Verlag Müller, Villingen/Schwarzwald, Germany, 1979 and 1978.

Blackwell, Dana. *Vienna Regulators of Lenzkirch and Lorenz Bob.* American Clock and Watch Museum, Inc. Bristol, CT. 1981.

The Jewelers Circular and Horological Review and *The Watchmaker and Jeweler* from 1875, 1884, 1886, 1888, 1891, and 1892 provided by Horological Data Bank National Association of Watch and Clock Collectors Museum, Inc.

Kochmann, Karl. *Black Forest Clockmaker and the Cuckoo Clock.* Antique Clocks Publishing, 1990.

_____. *Black Forest Music Clocks.* Antique Clocks Publishing, 1990.

_____. *Clock and Watch Trademark Index.* Antique Clocks Publishing, 1988.

_____. *Hamburg American Clock Company.* Antique Clocks Publishing, 1980.

_____. *Junghans Story.* Antique Clocks Publishing. 1976.

_____. *The Lenzkirch-Winterhalder & Hofmeier Clocks.* Antique Clocks Publishing. 1984.

Mühe, Dr. R. and H. Kahlert. *Furtwangen Beiträge zur Uhrengeschichte,* Band I, Uhren 1913.

_____. *Die Geschichte der Uhr.* Deutsches Uhrenmuseum Furtwangen, Callwey Verlag, München, Germany.

"Perpetual Calendar Clocks; American versus European." *Clockwise Magazine,* June 1980. Volume I, Number 11.

St. Louis Clock and Silverware Company. *1904 Twelfth Annual Catalog.* 1904.

Schaaf, Berthold. *Schwarzwalduhren.* Freiburg in Breisgau. Verlag Karl Schillinger, Germany, 1983.

Schneider, Wilhelm. "Frühe Kuckucksuhren von Johann Baptist Beha aus Eisenbach im Hochschwartzwald." *Alte Uhren und Moderne Zeitmessung,* June 3, 1987.

_____. "Die eiserne Kuckuckuhr." *Alte Uhren und Moderne Zeitmessung,* 5 October 1989.

Schneider, Wilhelm and Monika. "Black Forest Cuckoo Clocks at the Exhibitions in Philadelphia 1876 and Chicago 1893." *NAWCC Bulletin,* April, 1988, Volume 30/2, number 253.

Tyler, E. John. *Black Forest Clocks.* London, England, 1977.

Index

AF stamp on movement, 6

Badenia, 135
Badenia Black Forest Clock Manufactury, 30
Badische Uhrenfabrik, 30, 124, 127, 128, 130 *
Baduf, 127
Baeuerle, Tobias, 135
Bärmann, Thomas, 27
Bäuerle, Christian, 135
, Fridolin, 135
, Jacob, 198-200
, Karl, 198
, Mathias, 132
, Tobias, 135
Becker, Gustav, 132, 157
Beha, Engelbert, 107, 109, 111
, Johann B., & Söhne, 109
, Johann Baptist, 19, 25, 28, 29, 32, 33, 36, 53, 60, 84, 88, 95, 96, 100-122
, Lorenz, 107, 109, 111
, Martin, 100
, Vinzenz, 100
Blessing, Ernst, 139
, Karl, 235
Bob, Michael, 132
, Lorenz, 102, 111, 130
, Victor, 130
Braukmann, Hermann, 137
Bruder, Andreas, 233
, Gebrüder, 234
, Ignaz, 14, 233
, Ignatz & Söhne, 234, 235
, Wilhelm & Söhne, 234
, Xavier, 233, 234
Brugger Family, 100
Bürk, Johannes, 157
Burger, Christian, 22

Crisp Brothers, 243

Deurer, Wilhelm, 159
Dilger, Friedrich, 11
, J., 22
, Johann Friedrich, 13
, Mathias, 22
, Michael, 13
, Simon, 11
Dold, Adolf, 175
, Alfred, 175
, Carl, 22
, Josef, 175
, Karl Josef, 175
Dorer, Michael, 13, 16
Ducommun, David, 104
, Frederic-Guillaume Willian, 104
, Girod, 104, 105
, Henry, 104
, Jean, 104
, Louis, 104
Dufner, Johann, 11

Emmler, Erhard, 132

Faller, Am Bach, Leo, 124, 127
, Franz Josef, 169, 170, 173
, Friedrich, 124
, Mathias, 13
Fehrenbach, Adam, 124
, Albert, 132
, Rudolf, 129
Fellheimer, Jakob, 129
Firma Wintermantel & Co., 132
Frey, Lorenz, 11
Furtwängler, Gustav Adolf, 123
, Johannes, 123
, Julius Theophil, 123
, Karl Hektor, 123
, L., & Söhne (LFS), 109, 123-129, 173
, Lorenz, 123
, Oskar, 123
, Professor Adolf, 123
, Professor Wilhelm, 123

Ganter, Anton, 27
, Conrad, 100
, Josef, 13, 27

Gerwig, Robert, 169, 261
Gfell, Georg, 22
Goede, Franz, 159
Grance, Aaron D., 132
Grieshaber, Mathias, 19
Gunsser, Paul, 159

Haas, Karl, 132
, Karl & Gustav & Ludwig & Albert, 132
, Ludwig, 132
, Philip, and Sohne, 22, 30, 109, 132-136, 238
Hahn, Wilhelm, 50
Haller, Thomas, 109, 153, 156
, Jakob, 143
, Johannes, 143
Hamburg American Clock Co., 151, 157-159, 173
Harder, Anton, 132
Hauffe, Eugen, 47
Hauser, Eduard, 166, 169, 173
, Karl August, 169
, Paul Emil, 169
Hepting, Fidel, 53
Herbstreith, Jakob, 229, 230, 261
Hettich, Gordian, Sohn, 102, 109, 127, 129, 130, 131
, Hermann, 127
Hilser, Andreas, 109, 175
, Raimund, 175
Höfler, Felix, 100
Hofmeier, Johannes, 161-163
Hogg, Johann, 102
Holbfass, Adolf, 158
Hub, Friedrich, 49
Hummel, Adolf, 47
, Matthäus, 13, 26

Jäck, Father, 15
Jäger, Emil, 47
Jahresuhrenfabrik GmbH (August Schatz & Sohne), 109, 132
Jahresuhrenfabrik GmbH (Triberg), 132
Jerger, Wilhelm, 141
Jerome, Chauncy, 30, 250
Junghans, Arthur, 147, 151, 153
, Erhardt, 30, 147, 149, 151, 157
, Franz Xavier, 30, 147, 149, 151
, Frieda, 147
, Gebrüder, 109, 147-157, 159, 173
, Jr., Erhard, 151
, Nicholas, 147

Kaiser, Franz, & Oskar & Rudolf, 141
, Josef, 141
Kammerer, Josef, 13, 27
and Kuss, 73, 101
Ketterer, Anton, 107
, B., & Sons, 124
, Felix, 124
, Franz, 11
, Franz Anton, 13, 15, 53
, Theodor, 53, 107
Kienzle, Christian, 147
, Herbert, 147
, Jakob, 109, 145-147, 157
Kienzler, Augustin, 22
, German, 132
, Karl, 132
Kirner Family, 100
, Johann Baptist, 198
Koch, Professor Johann, 38, 47
Kreutye Brothers, 11
Kreuz, Paul, 22
Kuner, Gebrüder, 251-255

Lamy, Carl Konstantin, 261
Sons, 261
Landenberger, Christian, 158
, Paul, 151, 157-159
, Jr., Paul, 159
Landenberger & Lang, 151, 158
, Kurt, 159
, Richard, 159
Lang, Philip, 158
Lenzkirch-Aktiengesellschaft für

Uhrenfabrikation, 44, 45, 109, 157, 166, 169-197
Löffler, Mathias, 11
Lovell, G.S., 109, 111

Maier Albert, 135
, Andreas, 135
, Andreas (son), 135
, August, 135
, Brothers, 141
Martin, Jess Hans, 169
, Mathias, 233
Maurer, Rupert, 100
Maurer & Höfler, 109
Mauthe, Christian, 143
, Friedrich, 109, 143-145, 157
, Jakob, 145
, Johannes, 145
Mayer, Alois, 123
, Andreas, 123
, Frans-Karl, 123
, German, 123
, Heinrich, 123
, Joseph, 123
, Konrad, 123
, Theresia, 123
Mellert, Frans, 100
Merzbach, Lang & Fellheimer, 129
Meyer, Ferdinand, 137
Mittendorf, Louis, 104
Müller, Christian, 135
Munzer, Josef, 47

Noch, Heinrich, 137
, Johann Nepomuk, 137
Noll, August, 139, 142

Peerless, 135, 136
Pelhum, Charles E., 86
Pfaff, Maurer, 141

Reich Family, 100
Resch, Gebrüder, 153
Roder, Max, 129
Rogy, Nikolas, 169
Rombach, August, 124
, Carl, 124, 127
, J & C, 124
, Johann, 124

, Josef, 100
, Philemon, 47
Ruth & Söhne, 234
, Andreas, 234

Sattele, Matthä, 102
Scheffel, Viktor von, 8, 198
Scherzinger, F.X., 235
Schlenker, Carl Johannes, 145, 147
, Christian, 145
, E.R., 147
, Johannes, 145
Schöpperle, Ignaz, 166
, Johann George, 166
, Joseph, 132
Schwarz, Josef, 100
Schwer, August, 109
Schyle, Bertold, 22
Sears Roebuck & Co., Germany, 98
Sorg, Josef, 231, 232
Stegerer Family, 100
Stehling, Georg, 124
Steinbeiss, Ferdinand von, 149
Streifer, Jakob, 16, 234
Streyer, 15

Terry, Eli, 30, 151
, Samuel, 132
Teutonia Clock Manufactory, 30, 136
Thomas, Seth, 30, 151
Trenkle, Felix, 129
Tritscheller, Albert, 169
, Augustin, 109
, Johann Nikolaus, 169
, Paul, 169, 170, 201

Uhrenfabrik GmbH, A.G., Kienzle, (Schwenningen), 137
, Furderer Jaegler & Cie, 160-162
, Karl Josef Dold Söhne, 109, 175
, Neustadt, Schw. Aktiengesellschaft, 160
, Schwenninger, Schlenker & Kienzle, 137, 145, 147
, Villingen J. Kaiser, 141
, Villingen Aktiengesselschaft, 141
, Wehrle, 175
Union Clock Company, 30, 129
United Clock Factories, Junghans &

Thomas Haller, 156

Villing & Trenkle, 129
Villing & Fehrenbach, 129
, Josef, 129

Wehrle, Carl Joseph, 175
, Carl, 175
, Christian, 13
, Emilian, & Co., 41, 44, 48, 109, 198, 200, 201, 202-224
, Franz, 175
, F.X., 198
, Julian, 201
, Karl Raimund, 175
, Michael, 198
, Peter, 100, 102
Weisser, August, 129
Werner, C., 109
, C., Uhrenfabrikation, 137
, Carl, 137
, Hermann, 137
Wiest, Joseph, 169
Wilde, Alfred, 139
, Constantin, 137
, Gebrüder, 109, 137-139, 141
, Leopold, 137
Wildi, Johann, 100
Willman, A., & Co., 132
Wilmann, Georg, 11
Winstead Clock Factory, 151
Winterhalder & Hofmeier, 100, 109, 160-168, 173
Winterhalder, Anton, 163, 165
, Bernhard, 165
, George, 160
, Hermann, 165
, Johannes, 165
, Karl, 163
, Linus, 163, 165
, Ludwig, 163, 165
, Matthäus, 160-163
, Matthias, 160
, Thomas, 160, 163
Winterhalter, Jakob, 27
Wintermantel, Gerson, 132

Zeller, Jacob, 149

Price Guide

Rick Ortenburger

The demand for Black Forest clocks of all styles has risen dramatically in the last ten years. What used to be considered *junk* and hidden under tables at clock conventions is now highly collectable and you're lucky to find these clocks at conventions throughout the U.S.A. and Europe. Collectors throughout the world are looking for clocks similar to those pictured in this book. Because of this demand, this price guide should be used only as a guide and should not be the only determining factor in pricing these clocks.

The following should be taken into consideration when arriving at prices for these styles of clocks: Is the clock early or late? 30 hour or 8 day movement? Ornately or simply carved? Mint or busted/broken condition? In the case of a trumpeter, a two note or nine horn musical movement? Is the animated figure original and not repainted? Is the cuckoo bird original or plastic replacement? Etc.

The price range covered in this price guide reflects what the consumer might pay for a clock at an auction, flea market, or at the National Association of Watch and Clock Collectors convention in the United States, bought in *as is* condition to *restored* condition paid at a clock shop, or bought from a private individual. I arrived at these price ranges by collecting and selling these Black Forest clocks for fifteen years, and by also attending N.A.W.C.C. conventions and auctions throughout the United States and Europe.

This price guide is arranged by page number and clock position, and is listed in U.S. dollars. The position codes are as follows:

L: left
R: right
TR: top right
BR: bottom right
TL: top left
BL: bottom left
C: center
T: top
B: bottom
ALL: all clocks on page

13L	4000-8000	47L	150-400	74R	1000-2000
14L	8000-10000	47R	800-1500	75TL	1000-2000
14R	2000-6000	49L	200-400	75BL	300-600
16L	1500-4000	49R	300-750	75R	300-600
16R	3000-6000	50R	800-1200	76TL	300-600
17L	Unique	52L	600-1500	76BL	200-400
17TR	1500-4000	52R	100-350	76R	400-850
17BR	Unique	54R	4000-7000	77L	350-700
18L	3000-7000	56TL	250-750	77R	300-700
18R	500-1500	56BL	400-800	78L	500-1000
19R	500-1000	56R	550-850	78BR	1000-1500
20L	Unique	57L	800-1500	79TL	350-750
20TR	Unique	57TR	300-700	79BL	450-800
20BR	7000-10000	57BR	300-800	79R	450-900
21L	1500-4000	58TL	350-800	81TL	400-800
21TR	1000-2000	58C	300-700	81BL	700-1000
21BR	300-800	58TR	400-800	81R	1000-1500
22L	1000-3000	59TL	800-1500	82TL	400-850
24L	1000-4000	59BL	300-700	82BL	300-700
24TR	500-1500	60R	400-900	82R	200-400
24BR	800-1800	61TR	400-800	83L	500-1000
25L	2500-4000	61L	400-800	83TR	500-900
26L	3000-6000	61BR	450-900	83BR	600-1000
28TL	2000-4000	62L	300-600	84L	1000-1800
32L	2000-4000	62TR	300-800	84R	800-1200
33L	400-900	63R	700-1200	85L	600-1000
36TL	2000-3000	64L	300-800	85TR	600-1000
36BL	500-1500	65L	300-600	85BR	300-600
37TL	500-1200	65R	400-700	86L	350-750
37BL	500-1200	66L	700-1200	86TR	500-900
37R	500-1200	67TL	500-1000	87TL	400-700
39L	400-1000	67BL	1500-3000	100R	800-1200
39R	150-400	67TR	300-700	101TL	500-1000
40TR	1500-2500	68L	800-1500	101BL	1000-1500
41TL	3000-5000	69TL	800-2000	101R	1000-1500
41R	5000-9000	69BR	1000-2000	102TL	700-1300
42L	500-1500	70TL	300-800	102BL	500-1000
42R	150-400	70TR	1200-2000	103TL	400-800
43L	150-400	71L	1200-2000	104TL	2500-4000
43R	150-400	71TR	1500-2000	105TR	500-1000
44TL	2500-5000	72TL	500-900	105BR	1000-1500
44BL	1500-3000	72BL	500-900	106L	3000-4000
45L	3000-5000	72R	500-900	107R	400-800
45TR	750-1200	73L	1500-2000	108L	2000-3000
45BR	750-1200	73BR	500-900	109R	400-800
46L	200-400	74TL	400-800	110L	400-800
46R	150-400	74BL	400-800	111R	1200-1600

ID	Range	ID	Range	ID	Range
112L	400-800	171L	2500-3500	201BR	5000-8000
112TR	1000-1800	171BR	1500-2500	202TL	5000-8000
112BR	500-900	172L	600-1000	203BL	5000-8000
113L	800-1500	172R	5000-7500	204L	3000-6000
113TR	1000-1800	173TR	1500-2500	206L	6000-10000
114TL	1200-1800	173BR	1000-2000	207BR	4000-6000
115TL	1000-1500	174L	450-1200	208L	4000-6000
115BL	800-1200	175R	2000-4000	209R	7000-9000
115R	1200-1800	176L	900-1500	210R	2000-4000
116L	600-1000	176TR	600-1000	211TR	4000-7000
116TR	600-1000	176BR	300-500	212L	5000-8000
117TL	400-800	177TL	500-900	212TR	4000-6000
117BL	800-1500	177BL	100-200	214TL	6000-10000
117R	1000-1600	177R	400-900	215TL	4000-6000
118L	1000-1800	178TL	800-1500	215TR	5000-7000
121TL	400-800	178BL	800-1200	217L	1500-2500
121BR	400-800	178R	600-1000	218L	7000-10000
125L	1400-2500	179L	350-700	219L	3000-6000
126L	3000-6000	179R	1500-2500	222BR	5000-8000
127L	1500-2200	180L	1500-3000	223L	Unique
127R	1500-3000	180TR	1000-1500	224BL	4000-6000
128L	750-1250	180BR	800-1200	224BR	7000-10000
128BR	600-1000	181TL	300-600	225L	5000-7000
129TR	800-1500	181BL	1500-2500	225R	1500-2500
130TL	200-400	181R	1500-2200	226L	700-1500
131L	1200-1800	182L	450-1200	228L	700-1500
133TL	1200-1800	182TR	1200-2000	229R	350-700
136L	1500-3000	182BR	1000-1500	230TL	300-600
138R	1400-2500	183TL	200-400	230R	300-600
140L	1500-2500	183BL	300-600	231BL	300-600
140R	1000-1500	183R	1500-2500	232TL	1000-2000
142TL	Unique	184L	400-600	232C	1000-2000
142TR	1500-2500	184TR	300-600	232TR	1000-2000
142BL	1000-2000	184BR	1000-1500	233R	3000-6000
143R	250-500	185TL	2000-4000	235L	8000-10000
144L	700-1000	185BL	300-500	235C	6000-8000
144R	400-800	185R	2000-3000	236R	2000-4000
145R	1500-2500	186L	1000-2000	238L	2000-4000
146TL	900-1500	186TR	300-500	238R	2000-4000
146R	350-1000	186BR	400-800	239R	1500-3000
147R	600-1000	187T	100-200	240L	600-1000
148L	750-1500	187B	1000-2000	241TL	350-700
148C	350-950	188L	800-1500	241R	350-700
148R	450-1200	188TR	2500-4000	242L	2000-4000
150L	300-850	188BR	350-750	242TR	600-1000
151L	800-1500	189TL	400-600	243L	3000-5000
151R	400-850	189BL	500-900	244L	1000-2000
152TL	400-800	189R	750-1500	244BR	1500-3000
152R	500-1000	190L	2000-4000	245TL	1000-1800
153TL	400-800	190TR	800-1500	245R	1500-2500
153BL	400-800	190BR	2000-3000	246L	1200-2200
153R	800-1500	191TL	800-1500	246R	1200-2200
154BL	150-250	191BL	800-1200	247TL	800-1500
154BR	150-250	191R	600-1000	247BL	1200-2200
155L	450-850	192L	700-1200	247R	1000-2000
155TR	400-1200	192C	1000-2000	248L	1000-1800
157R	400-1250	192R	600-1200	248C	400-950
158R	200-600	193L	1000-2000	248R	750-1500
159R	150-250	193C	850-1500	249L	700-1100
161TL	1200-2000	193R	800-1500	249C	2000-4000
163TL	200-400	194L	1000-2000	249R	800-1800
164TL	500-1000	194TR	800-1500	250L	1200-2000
164TR	500-1000	194BR	600-1000	250C	1200-1800
164BR	600-1000	195TL	800-1500	250R	1000-2000
165TL	600-1000	195BL	150-300	252All	200-500
165BL	400-700	195R	1500-3000	253All	400-800
165TR	500-900	196L	800-1500	254All	400-800
166L	600-1200	196R	500-900	256B	1500-3000
167L	3000-4500	197TL	800-1500	257C	2000-3000
167TR	800-1500	197BL	600-1000	258L	300-600
170L	300-600	197R	800-1500	258R	500-1000
170R	2000-3500	199TR	6000-10000		

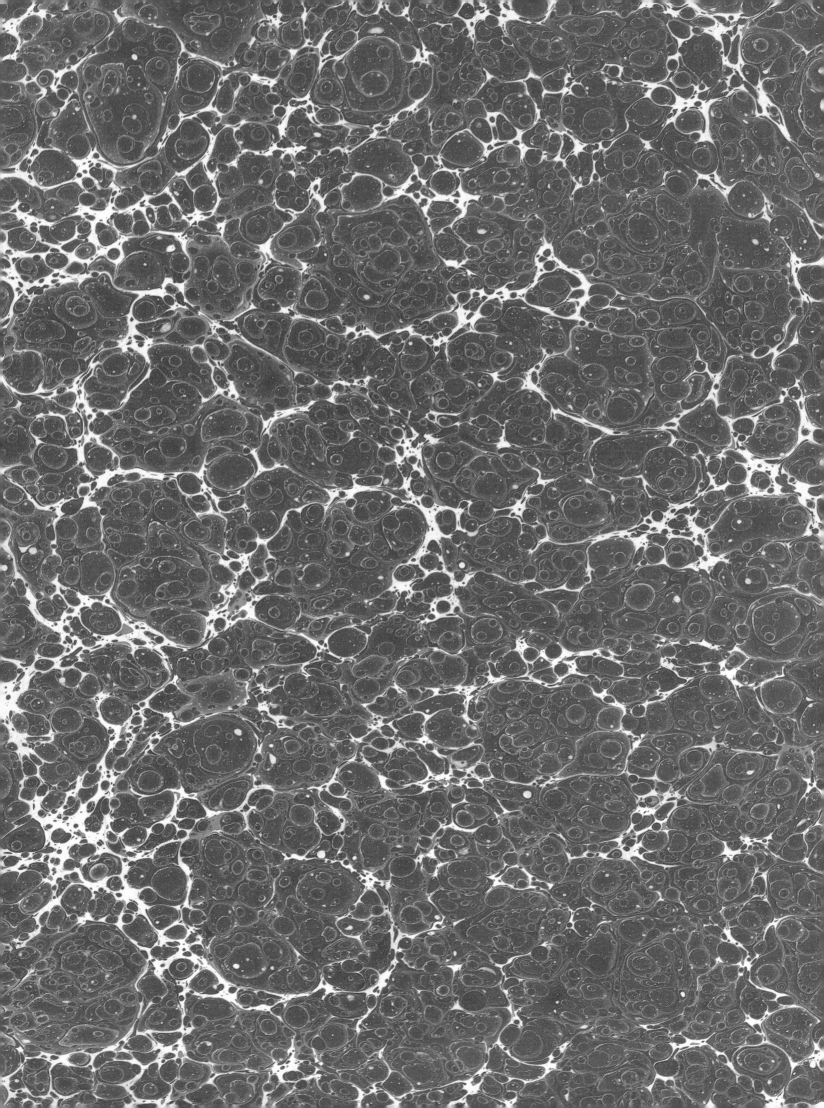